Tomorrow's Geography

for Edexcel GCSE Specification A Revision Guide

Mike Harcourt • Steph Warren

Hodder & Stoughton
A MEMBER OF THE HODDER HEADLINE GROUP

Cover images from Photodisk, photograph of Machu Picchu is taken by Tim Smith.

Every effort has been made to trace and acknowledge ownership of copyright. The publishers will be glad to make suitable arrangements with any copyright holders whom it has not been possible to contact.

Note about the Internet links in the book. The user should be aware that URLs or web addresses change regularly. Every effort has been made to ensure the accuracy of the URLs provided in this book on going to press. It is inevitable, however, that some will change. It is sometimes possible to find a relocated web page, by just typing in the address of the home page for a website in the URL window of your browser.

Orders: please contact Bookpoint Ltd, 130 Milton Park, Abingdon, Oxon OX14 4SB. Telephone: (44) 01235 827720. Fax: (44) 01235 400454. Lines are open from 9.00–6.00, Monday to Saturday, with a 24 hour message answering service. You can also order through our website www.hodderheadline.co.uk.

British Library Cataloguing in Publication Data
A catalogue record for this title is available from the British Library

ISBN 0 340 811 08 0

First Published 2003
Impression number 10 9 8 7 6 5 4 3 2 1
Year 2009 2008 2007 2006 2005 2004 2003

Typeset by Pantek Arts Ltd, Maidstone, Kent.
Printed in Great Britain for Hodder & Stoughton Educational, a division of Hodder Headline Plc, 338 Euston Road, London NW1 3BH

Contents

Introduction

This revision guide is designed to help you to prepare for your GCSE Geography in the easiest way. It is specifically written for those of you who are studying Edexcel Specification A.

How to use this book

- All the information you need is presented in brief facts and explanations
- There are diagrams to help you to remember your work
- Some of the case studies have been put into mind maps. This should help you to revise in a visual way
- The first four chapters refer to the core units and you will need to learn all of this
- The next four chapters are the option units you may need to ask your teacher for help to determine the units you have studied
- Throughout the book there are smiley faces drawn, these tell you important facts about what examiners are looking for and the sort of questions to expect on the papers

Remember

The first paper, dealt with in chapters 1–4, will have an OS map and photographs. It is more a test of your geographical skills. The second paper, dealt with in chapters 5–8, is a test of your geographical knowledge. In both papers there are case studies that you need to learn.

1 The physical world

Processes which occur on coasts

Weathering – ways that rocks are broken down

Physical weathering: freeze–thaw action

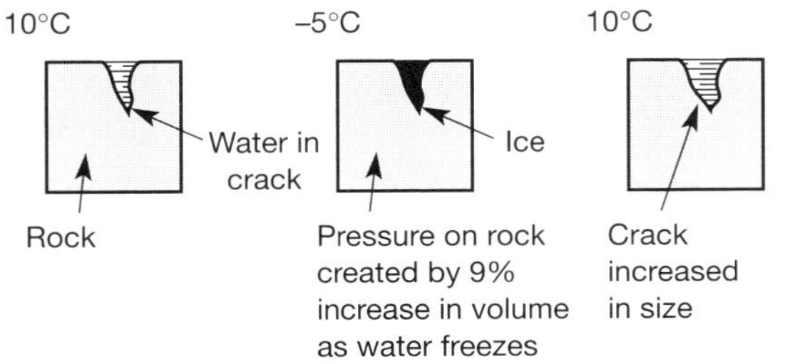

Physical weathering: onion skin

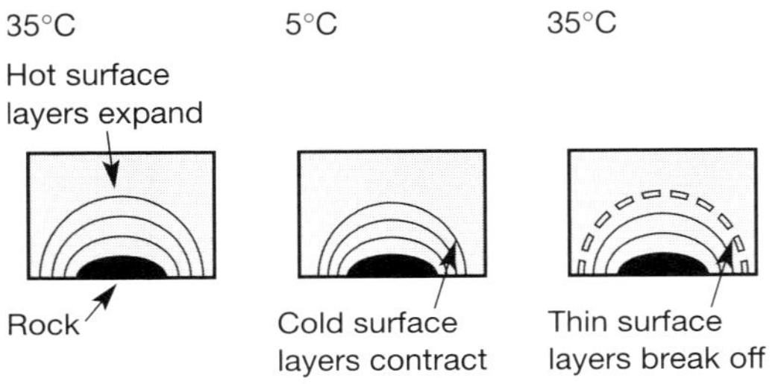

Biological weathering

Seed falls into crack

↓

Rain causes seedling to grow

↓

Roots force their way into cracks

↓

As the roots grow they break up the rock

Chemical weathering

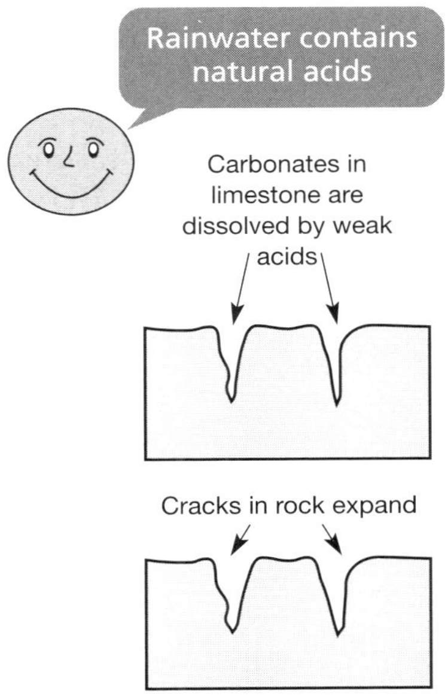

Mass movement – this is when material moves down a slope pulled by gravity

There are two main forms: soil creep and slumping.

Soil creep

- Slowest downhill movement
- Gravity pulls water in the soil downhill
- Soil particles move with the water
- Heavy rainfall causes faster downhill movement
- The slope appears to have ripples
- The ripples are known as terracettes

Slumping

- A large area of land moving down a slope
- Common on clay cliffs
- Dry weather makes the clay contract and crack
- When it rains water gets into the crack
- The soil becomes saturated
- A large piece of rock is pulled down the cliff face
- It has slipped on the slip plane of saturated rock

Erosional processes – these help to create coastal landforms

Corrasion – sand and pebbles carried in waves are thrown against the cliff face.
Corrosion – chemicals in sea water dissolve certain rock types e.g. chalk.
Hydraulic action – the compression of air in cracks which puts pressure on the rock and causes pieces of the rock to break off.

The following are processes that are happening in the sea:
Attrition – this is the breaking up of rocks and pebbles in the waves.
Transportation – this is the movement of sand and pebbles by the sea.
Deposition – this is the putting down of sand and pebbles by the sea.

Coastal landforms – the interaction of processes result in a range of coastal landforms

Erosional features – cliffs and wave-cut platforms

Below is an annotated diagram which explains the formation of cliffs and wave-cut platforms.

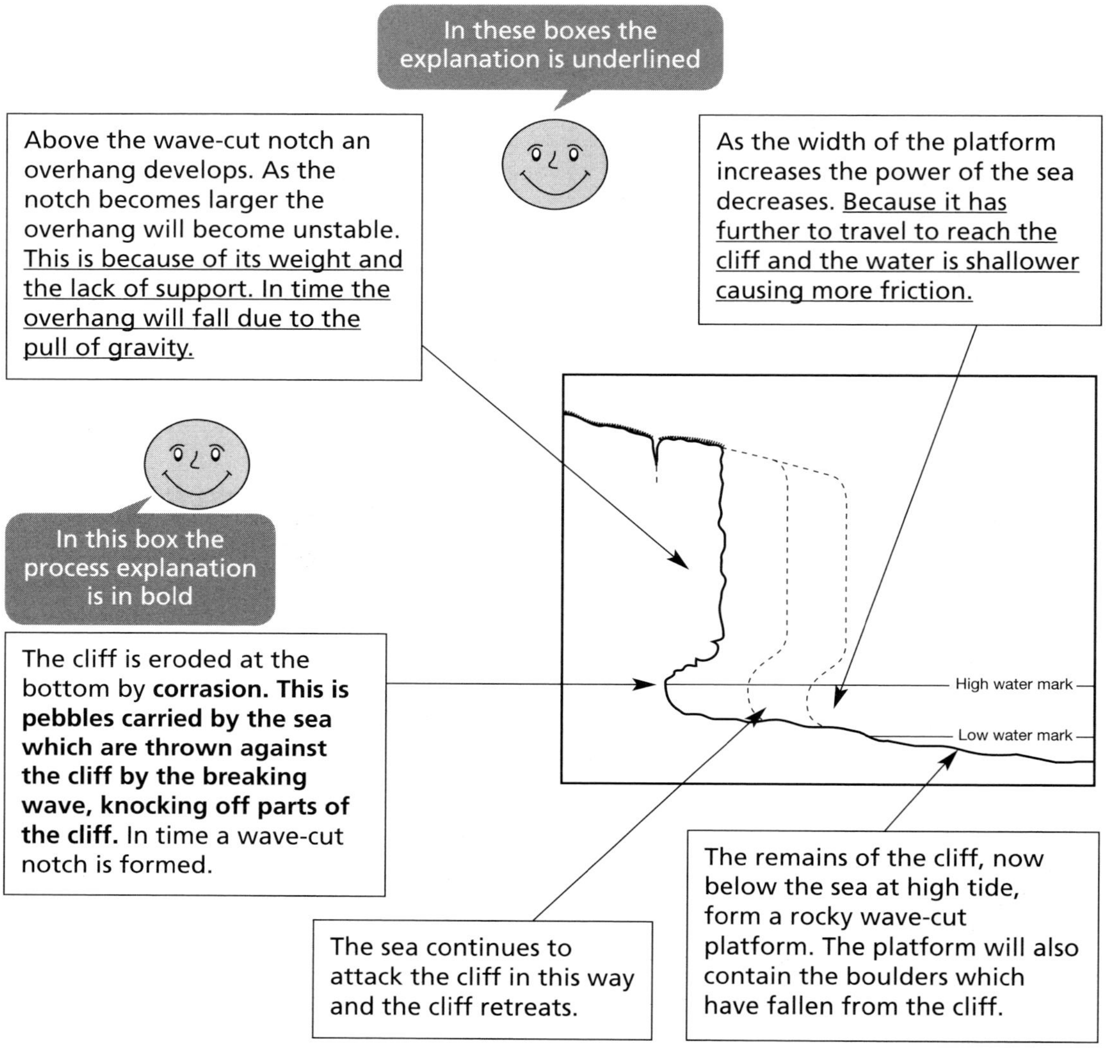

Erosional features – headlands and bays

These form due to different rock types.

- Only occur on coastlines where soft and hard rocks are found at right angles to the sea
- The soft rock erodes more quickly than the hard rock forming bays
- The hard rock is more resistant and sticks out as headlands
- There should also be reference to the processes of erosion
- A* grade candidates may also refer to the fact that erosion eventually becomes greater on the headlands because the bays have retreated and the headlands are more exposed. The sea is not as powerful in the bays because it is shallower and therefore starts to deposit material

Erosional features – caves, arches, stacks and stumps

How to answer questions which ask for description and explanation

Below describes how a stump is formed. There is no explanation given.

This is just description

A stump is formed by the action of the sea and weathering. The sea erodes a crack using hydraulic action and makes it bigger forming a cave. If it is a headland caves will form on either side. Eventually the backs of the caves meet and an arch is formed. In time the arch will collapse forming a stack. The stack will then collapse forming a stump.

Below is an A answer that explains how a stump is formed. Note the comments, which pick out the sequence description, explanation and the processes.*

Sequence description → A stump is formed by the action of the sea and weathering.

Process explanation → The sea erodes a crack with hydraulic action. This is when water hits the cliff compressing air in cracks. As the water retreats the pressure is released breaking off pieces of rock.

Sequence description → This makes the crack bigger forming a cave. If it is a headland, caves will form on either side. Eventually the backs of the caves meet and an arch is formed. In time the arch will collapse forming a stack.

Process explanation → This is due to undercutting of the sea. One of the processes the sea uses is corrasion which is when rocks in the sea are thrown against the cliff breaking off pieces of the cliff.

Sequence description → The arch becomes wider at the bottom and is unable to support the weight above. Eventually the arch is pulled down by gravity.

Sequence and process → Weathering is also active on the cliff face and further weakens the cliff. Through time the sea erodes the bottom of the stack and it collapses to leave a stump, which is covered by the sea at high tide.

This is a C grade answer. Again the sequence description, explanation and processes are picked out.

Process → A stack is formed by the action of the sea. The sea erodes a crack in the cliff using corrasion. Eventually the cave becomes an arch. In time the arch will collapse forming a stack. ← **Sequence described**

Sequence explained → This is due to undercutting of the sea. The arch becomes wider at the bottom and is unable to support the weight above. Eventually the arch is pulled down by gravity.

Read the question carefully. Have you been asked to describe or explain?

You must include sequence and processes on all landform formation questions

Don't forget a diagram will always help your answer

River landforms – the interaction of erosion and deposition forms a number of distinctive river landforms

Waterfall formation

1

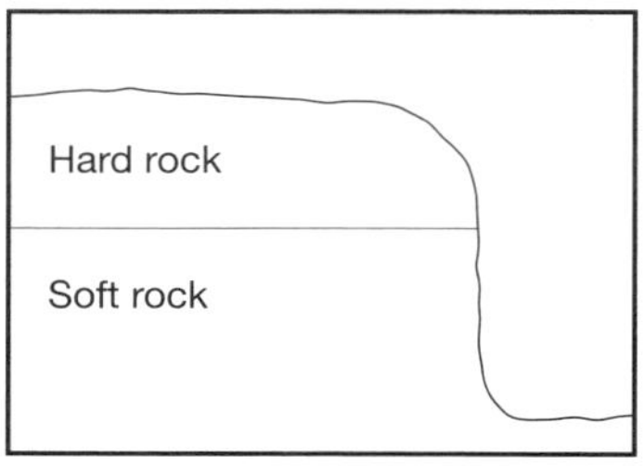

2

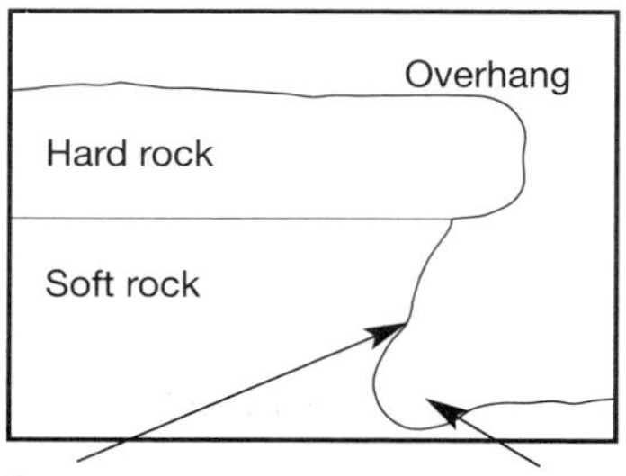

Soft rock eroded back more quickly (by splash back) caused by hydraulic action

Plunge pool formed by the force of the water

3

Overhang has fallen due to lack of support and pull of gravity

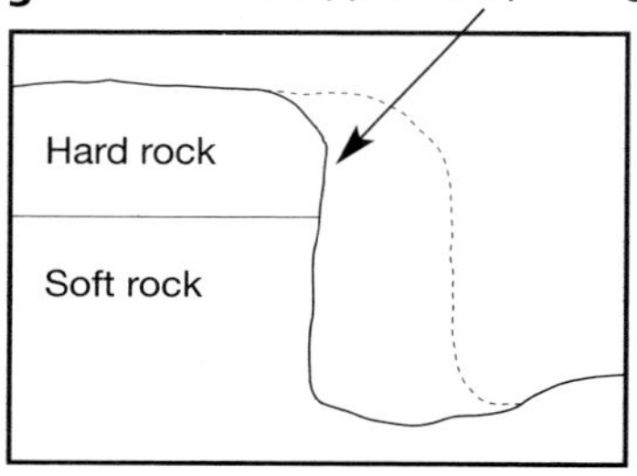

Waterfall retreats upstream to leave a gorge

River features and landforms

As a river moves downstream:

- Gradient decreases
- Becomes wider and deeper
- Velocity and volume increase

Erosional processes

These are similar to those attacking the coast. Hydraulic action, corrasion, corrosion and attrition.

Rivers also transport and deposit material along their courses.

Interlocking spurs

These are pieces of hard rock that the river winds round as it goes down the valley.

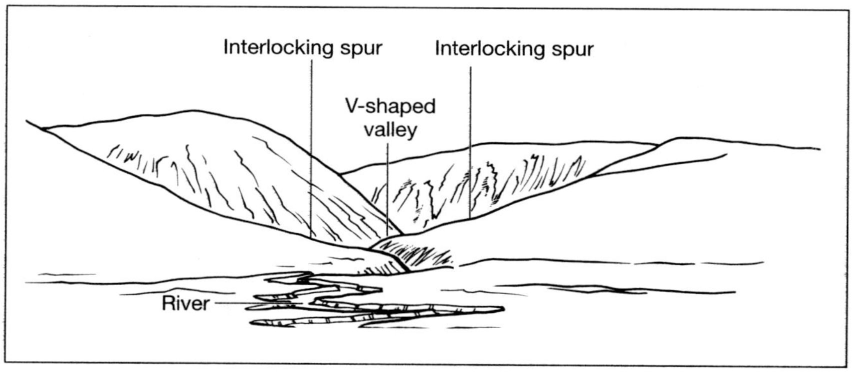

Meander bends and oxbow lakes

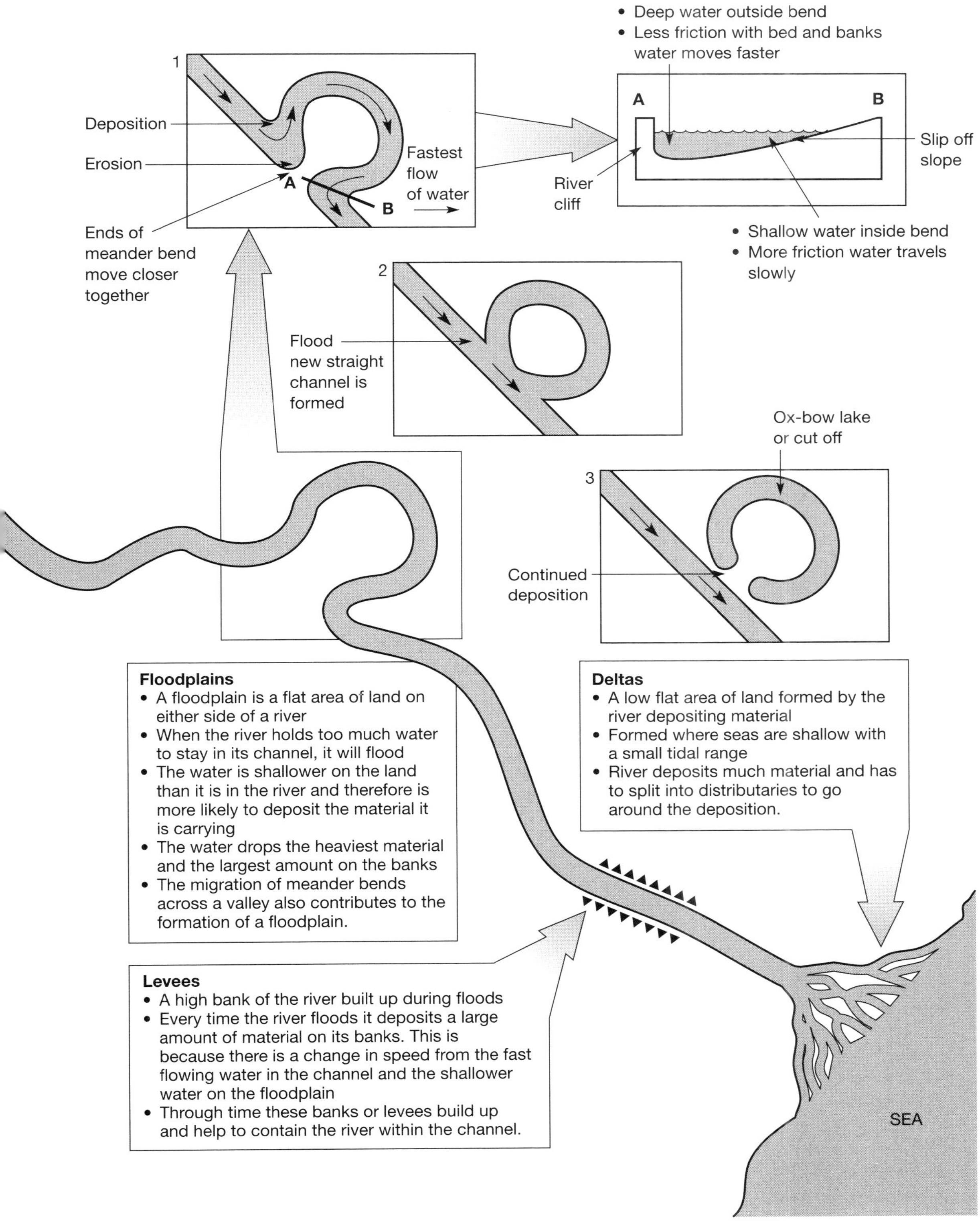

Floodplains
- A floodplain is a flat area of land on either side of a river
- When the river holds too much water to stay in its channel, it will flood
- The water is shallower on the land than it is in the river and therefore is more likely to deposit the material it is carrying
- The water drops the heaviest material and the largest amount on the banks
- The migration of meander bends across a valley also contributes to the formation of a floodplain.

Deltas
- A low flat area of land formed by the river depositing material
- Formed where seas are shallow with a small tidal range
- River deposits much material and has to split into distributaries to go around the deposition.

Levees
- A high bank of the river built up during floods
- Every time the river floods it deposits a large amount of material on its banks. This is because there is a change in speed from the fast flowing water in the channel and the shallower water on the floodplain
- Through time these banks or levees build up and help to contain the river within the channel.

Glacial processes and landforms

Corries

1.

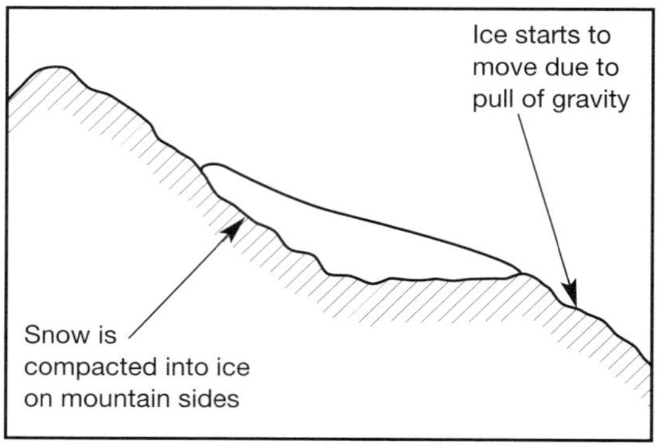

2.

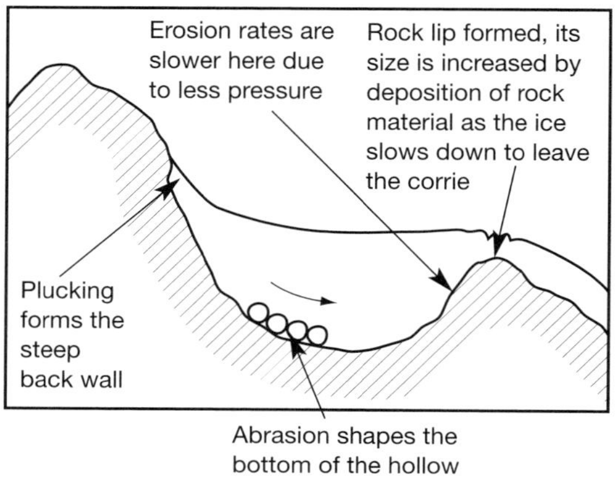

3.

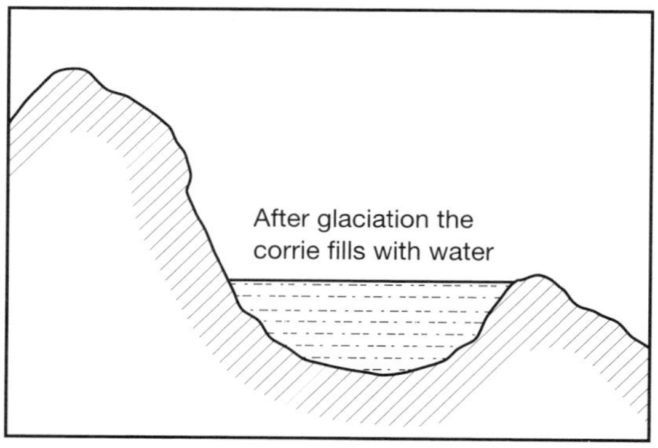

Arêtes and pyramidal peaks

- If two corries form beside each other on a mountainside their sides will become knife-edge ridges known as arêtes. These are continually attacked by freeze–thaw weathering
- If corries form on at least three sides of a mountain the top becomes a sharp peak of jagged rock known as a pyramidal peak

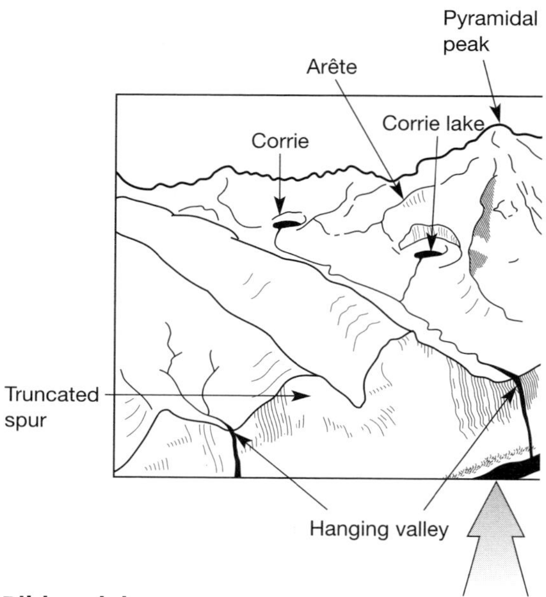

Ribbon lakes

- These are long narrow lakes in valleys in upland areas that have been glaciated
- They were formed when the glacier met less resistant rock on the valley floor and eroded more here to form a dip in the valley bottom
- After glaciation this filled with water

Glacial erosion processes – there are two types of glacial erosion

Plucking – this is ice from the glacier melting due to pressure, the water runs into cracks on the mountainside. It refreezes almost immediately breaking off pieces of rock, which become embedded in the glacier.

Abrasion – the rocks which are in the glacier are then used by the glacier to erode the sides and bottom of the valley as the glacier moves.

Before glaciation

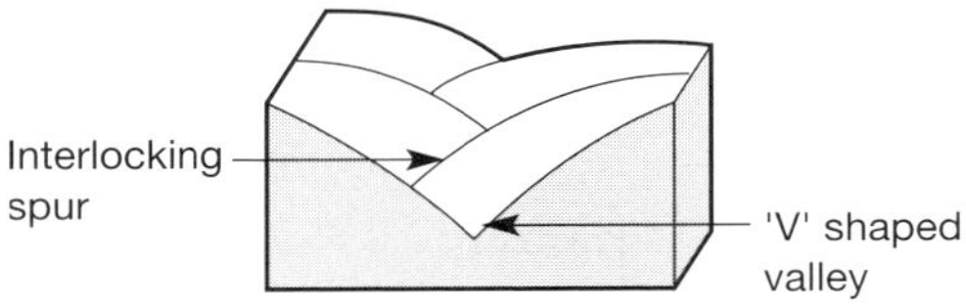

The diagrams show the same valley before and after glaciation

After glaciation

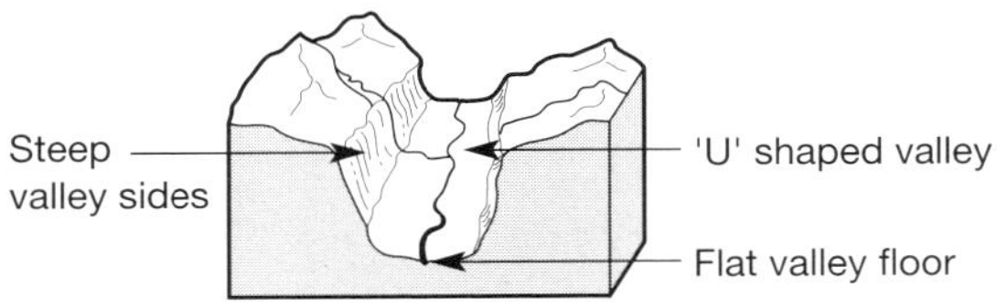

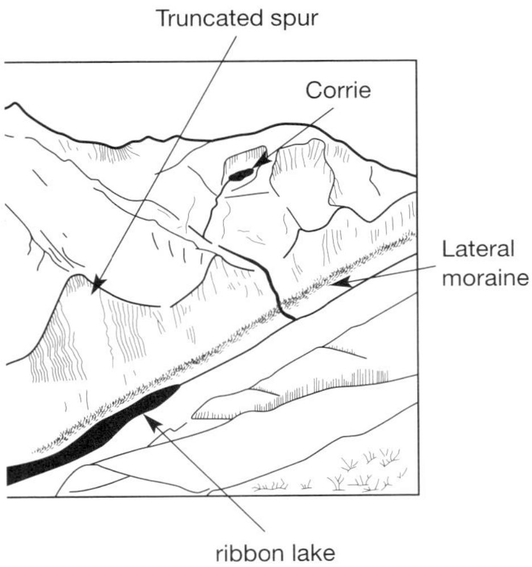

Remember! When answering a question on landform formation
Sequence
Process
Explain
Describe

U-shaped valleys and truncated spurs

- A valley glacier will completely fill a river valley. It has a lot of power to erode because of its size and weight
- This great power cuts back the interlocking spurs using abrasion and plucking forming truncated spurs
- This leaves a wide flat bottomed valley with steep cliff-like sides

Hanging valleys

- These are the valleys of tributary streams
- They are left high up on the valley sides of the main valley because their glaciers were smaller and had less erosive power than the main valley glaciers
- When the ice melted these valleys were at a higher level. These streams now meet the main river via a waterfall

Moraine

- This is material transported and deposited by glaciers
- There are four types of moraine
- Moraine is collected by the glacier from freeze–thaw weathering, plucking and abrasion
- Moraine will be dropped where the glacier slows down
- Lateral moraine is dropped at the sides of glaciers
- Ground moraine is put down below the glacier
- Where two glaciers meet their lateral moraines become medial moraines
- Terminal moraine is dropped where the glacier ends

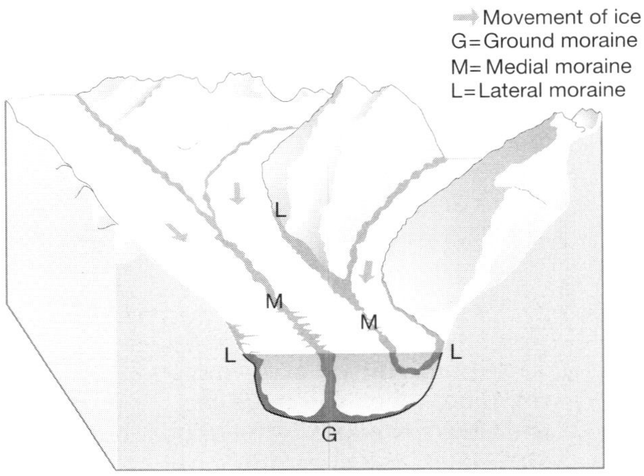

Depositional features – beaches

beaches

- found on straight coastlines where LSD is happening
- found in bays where the sea is shallower so the waves lose their energy and deposit what they are carrying
- made up of sand and pebbles
- formed by constructive waves

Depositional features – spits and bars

- Spits are a narrow stretch of sand and pebbles that are joined to the land at one end
- Bars are spits which go across a bay
- This is only possible if there is shallow water and no river entering the sea
- Therefore if you were asked to explain the formation of a bar you would have to discuss spit formation and then add on this bit

Below is an annotated diagram which explains the formation of a spit. There is also a useful diagram explaining longshore drift.

Useful diagram on LSD!

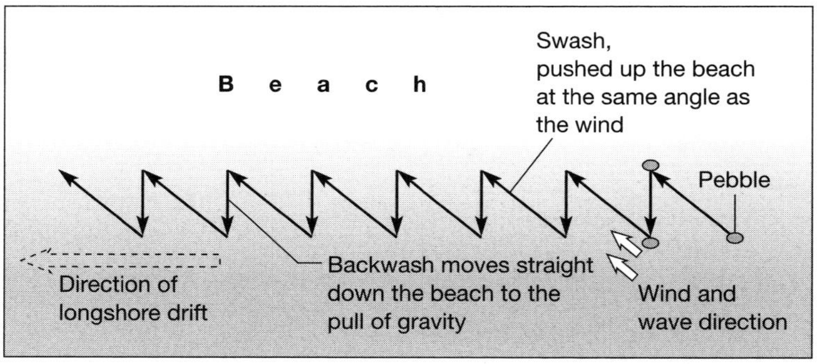

Descriptive

Material is moved along the coast because of longshore drift.

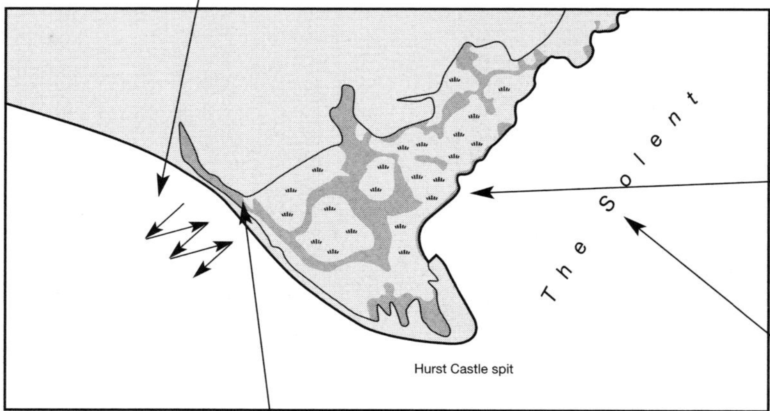

Descriptive

Salt marshes form behind the spit.

If longshore drift is occurring on a coastline. When the coastline ends, the sea deposits the larger material it is transporting in the sheltered water by the headland. This is due to a change in depth.

Sequence explained

Explanation again

If there is a river estuary then the meeting of the waves and the river water causes a change in speed and results in both the waves and the river dropping their sediment.

2 The human world

World population change

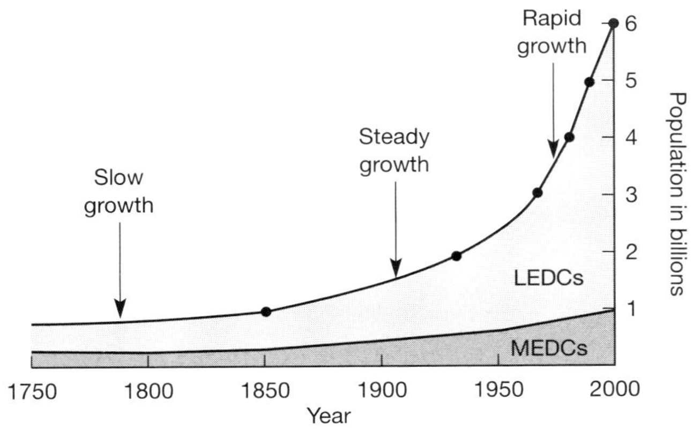

Population predictions

- Global population growth is now beginning to slow
- 95% of growth is in LEDCs
- 9 billion by 2050
- Population will reach maximum 10.4 billion in 2200

The balance between birth and death rates

World population growth is caused when birth rates are higher than death rates.

- Birth rate is the number of births per 1000 people per year
- Death rate is the number of deaths per 1000 people per year

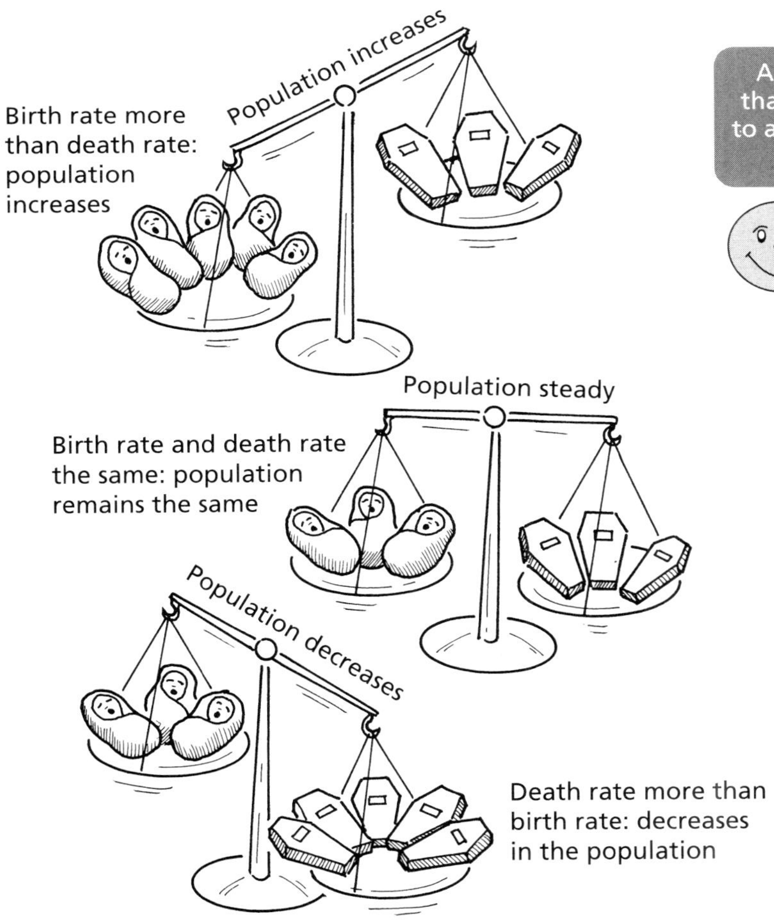

A birth rate higher than death rate leads to a natural increase in the population

Remember population doesn't always have to go up. It can also go down

Birth and death rates in LEDCs

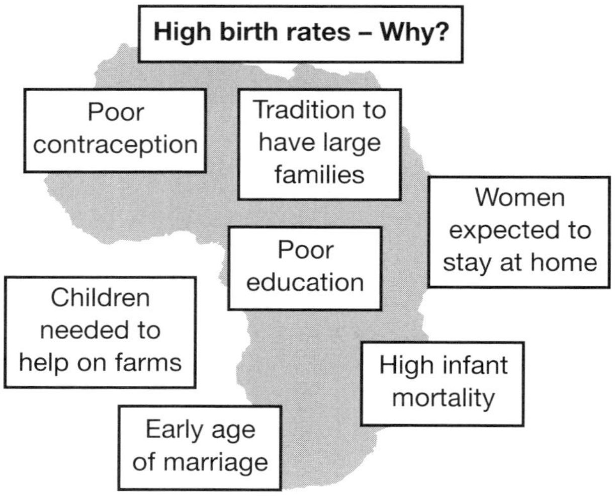

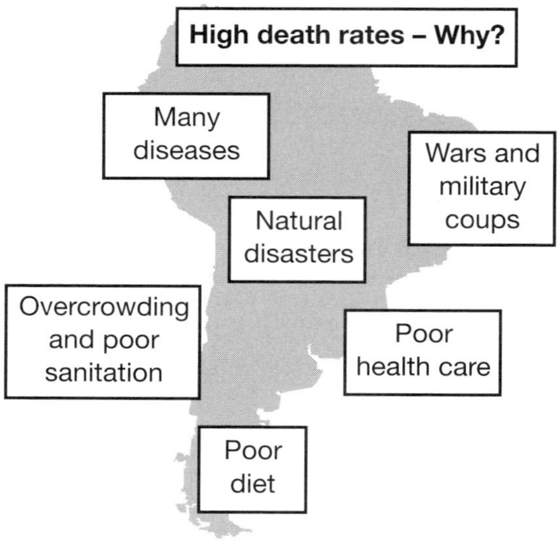

Birth and death rates in MEDCs

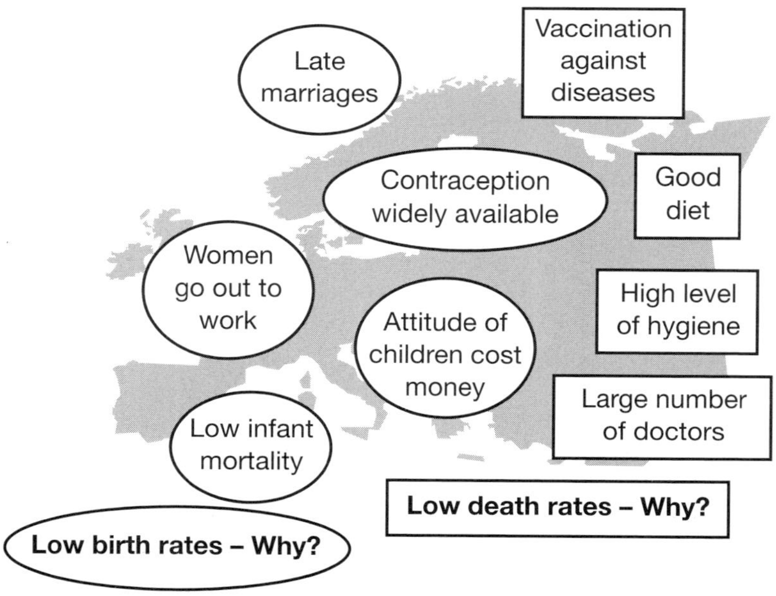

In the exam you might be given a country and asked to explain its birth and death rates

Demographic transition model

This model describes the way the total population of an area changes through time because of differences in birth and death rates.

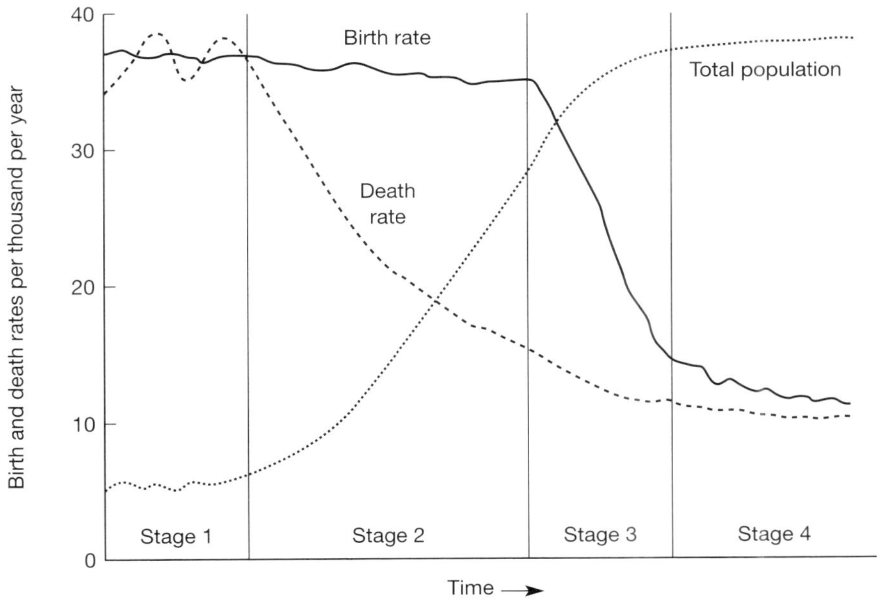

Stage	one	two	three	four
Birth rate	High	High	Falls	Low and fluctuates
Death rate	High and fluctuates	Falls	Low	Low
Population change	Small growth	Rapid growth	Slower growth	Stable
Places	Amazon tribes	Poor LEDCs e.g. Mali	Developing LEDCs e.g. Taiwan	MEDCs e.g. UK
UK	Before 1750	1750–1880	1880–1950	1950–present
Stage characteristics	Subsistence agriculture, high infant mortality	Improved food supply, no birth control, some medical advancement, little secondary education	Better living conditions, improving health care and hygiene, birth control, rapid urbanisation	High standards of living, education for all, high level of tertiary employment, late marriage

Many MEDCs may be entering a fifth stage where death rates are lower than birth rates causing the population to go down.

The model can be used to predict changes in population and future needs for education and health services.

Exam tip:
You may be asked to explain the birth and death rates in each stage

Migration

Migration is when people move from one place to another to live or work.

The number of people leaving or entering a country or an area will change its population.

Population pyramid for Calcutta

Large city in an LEDC showing the impact of rural to urban migration

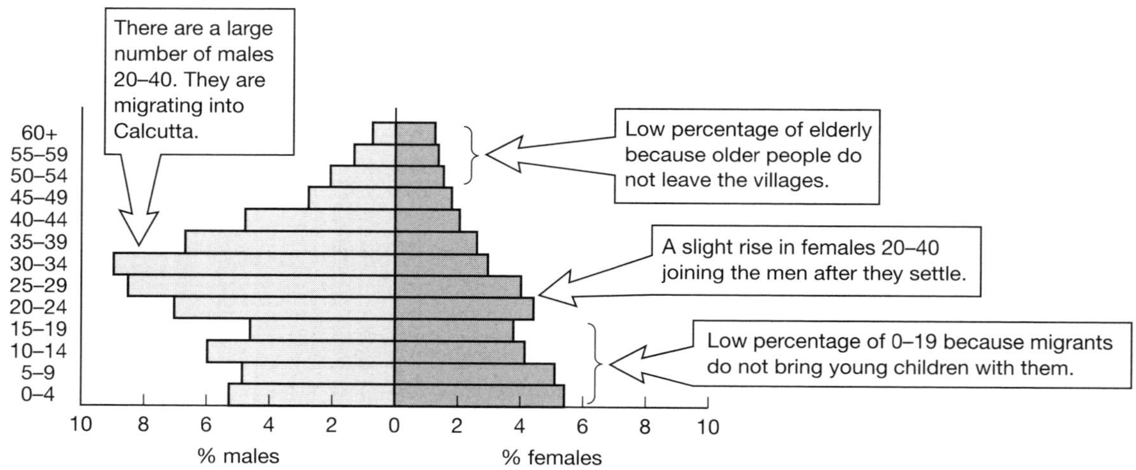

Why did birth rates fall in stage three?

- Contraception becomes common
- Laws make it illegal for young children to work
- People marry later
- Women go to university and have careers

Population pyramids

Population pyramids are a way of showing the age and sex of the people

You need to know the general shapes for LEDCs and MEDCs

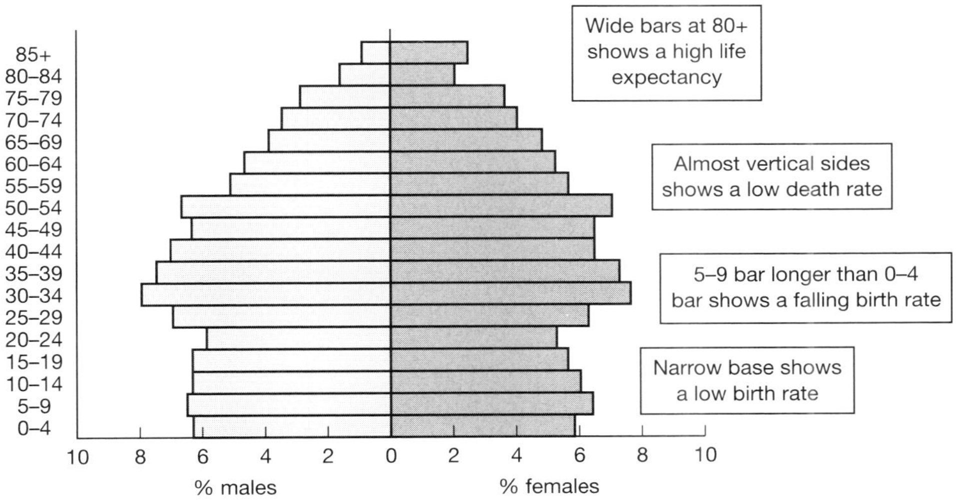

MEDC UK pyramid

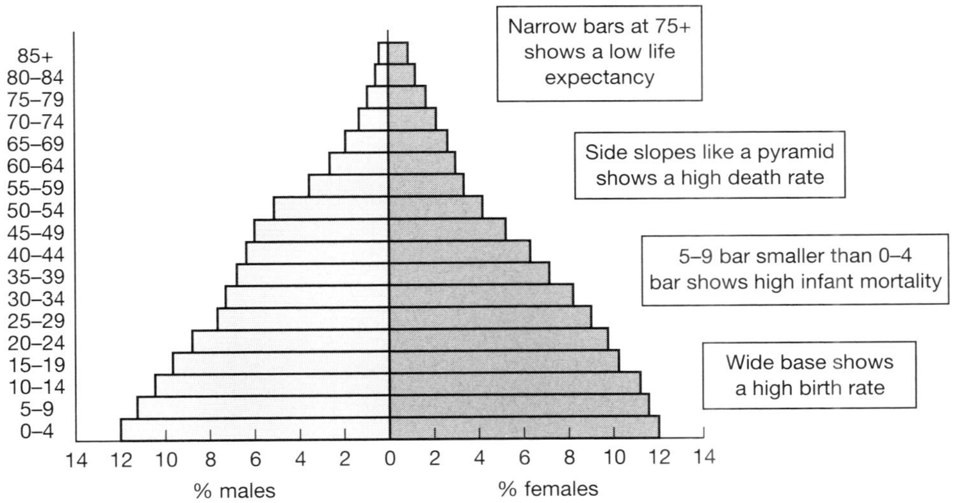

LEDC India Pyramid

In the exam you might be asked to construct part of a pyramid or to take information from the pyramid

Population structure

You need to know about age, sex, ethnic, religious and employment characteristics

How are these characteristics different between LEDCs and MEDCs?

Census

You can find out a lot about the population structure by looking at a census.

In the UK a census is carried out every 10 years. The first one was in 1801.

Comparing rural and urban populations

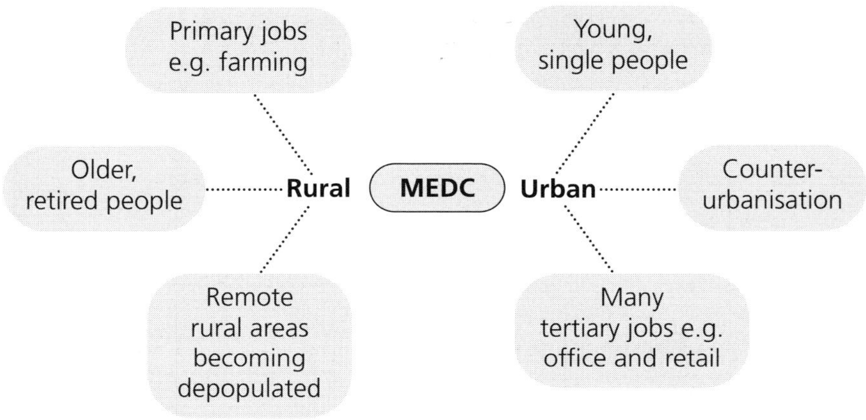

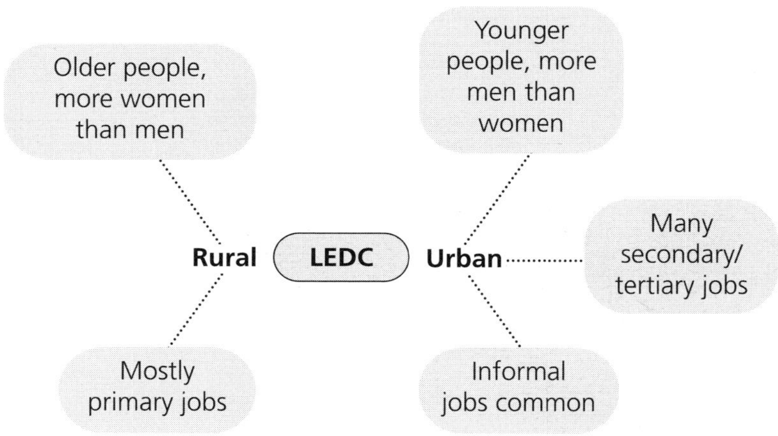

Dependency ratio

The young people (0–15) and the old people (65+) are supported financially by the economically active (people of working age 16–64). This relationship is called the dependency ratio.

It can be written as a formula.

$$\text{Dependency ratio} = \frac{\text{young (0–15)} + \text{old (65+)}}{\text{economically active (16–64)}}$$

> You may be asked to work out a dependency ratio. Can you?

- A dependency ratio of 1 means each dependant is supported by one worker
- Higher figures show more people are dependent on fewer workers
- LEDCs tend to have high figures because they have so many children
- Dependency ratios in MEDCs are getting higher as more people live longer

> People who don't work are financially dependent on the workers

What are the problems of an ageing population?

- Old people cost a lot to support
- Shortage of hospitals and care homes for the elderly
- Pension schemes under pressure
- Working population will have to pay more taxes
- Retirement age will have to go up
- All buildings will need to be 'old people friendly'

> An ageing population can also have some advantages
> - Many elderly people in MEDCs have money to spend which helps the economy
> - Charity work is mostly manned by retired people
> - The tourist industry has benefited with more retired people going on several holidays a year

What are the problems of a youthful population?

- Schools become overcrowded
- Not enough doctors and midwives trained to look after children
- Not enough workers to support dependent population
- More difficult for women to work

> A youthful population can have some advantages
> - Provides a steady supply of workers
> - Children provide security and support for old age

Settlements

What affects the location of settlements

Physical factors

Relief
• hills • valleys
• slopes

Water
• drainage • adequate supply for drinking

Climate
• rainfall • temperature
• wind • aspect

Soil
• fertility • depth
• drainage

Economic factors

Transport
• road • rail
• water

Mineral resources
• energy e.g. coal
• commercial e.g. gold
• power e.g. HEP

Employment
• ports • tourism
• mining • agricultural

You will be expected to use ordnance survey maps, sketch maps and photographs to identify locational factors

Physical factors in locating a settlement

Physical factors were very important when locating early settlements.

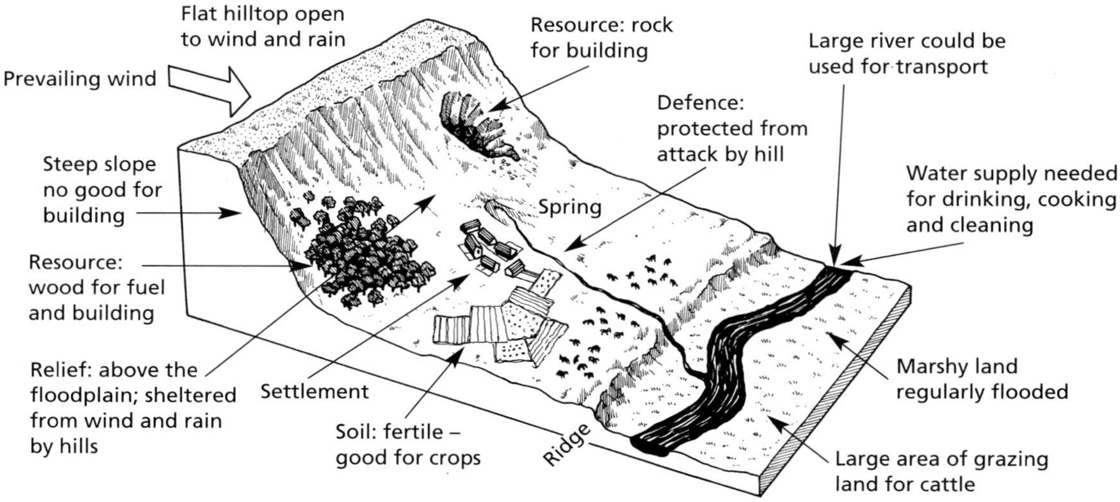

Today many of these reasons for the location of settlements are outdated. Economic factors have become more important.

Settlement shape

Settlements can be classified according to the arrangement of houses within them. Settlements may be linear (ribbon-like), nucleated (clustered), dispersed (fragmented), rectangular, circular or star-shaped.

The shape of the settlement is affected by

- Transport routes
- Natural barriers such as coastlines and lakes
- Relief such as steep slopes
- Planning controls such as green belts

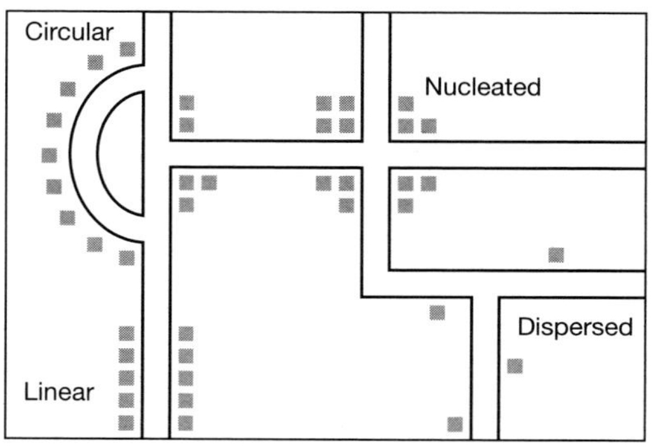

Land use in an urban area in an LEDC. Know your case study – Bangalore, India

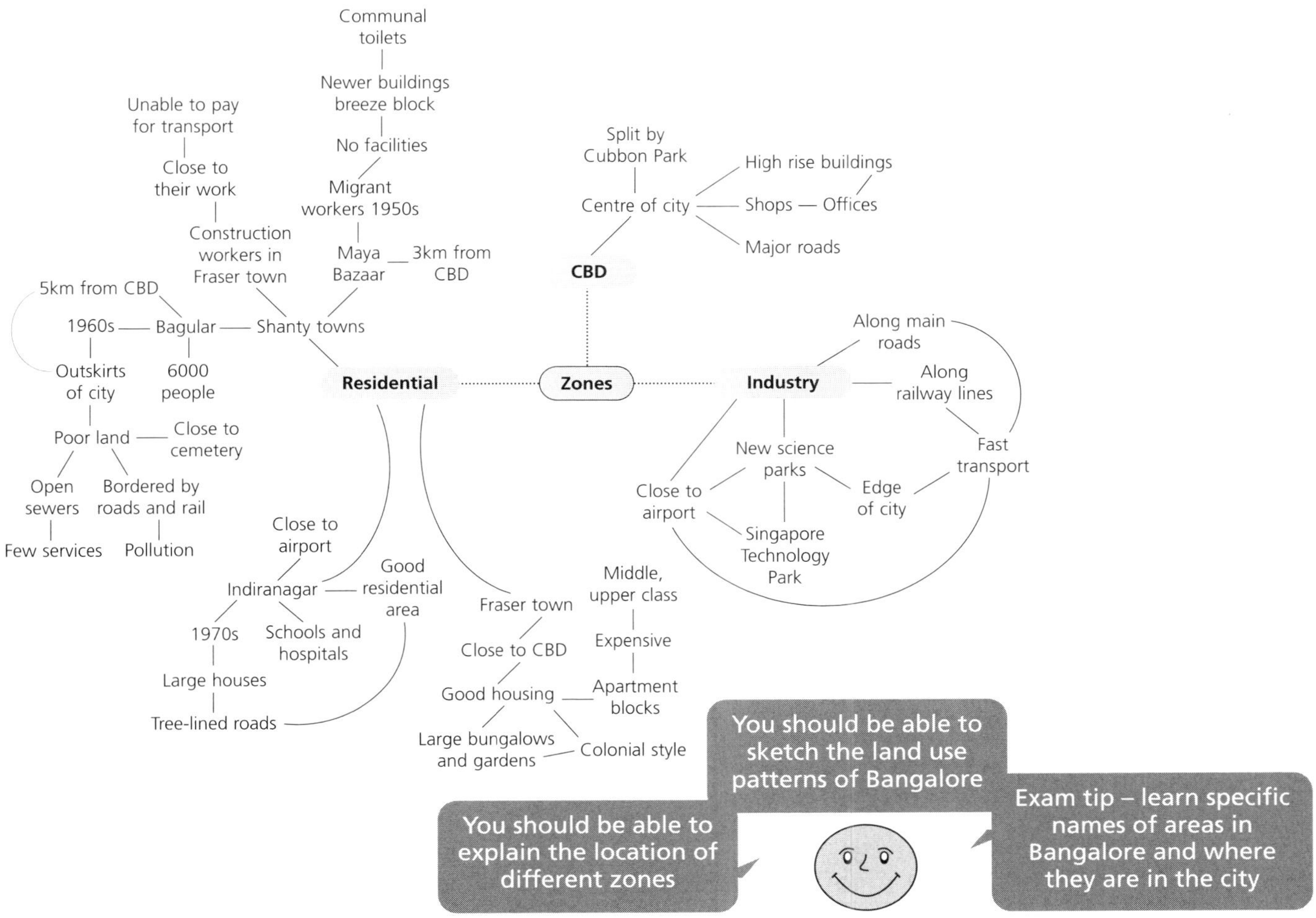

Land use in an urban area in an MEDC. Know you case study – Reading, Berkshire, UK

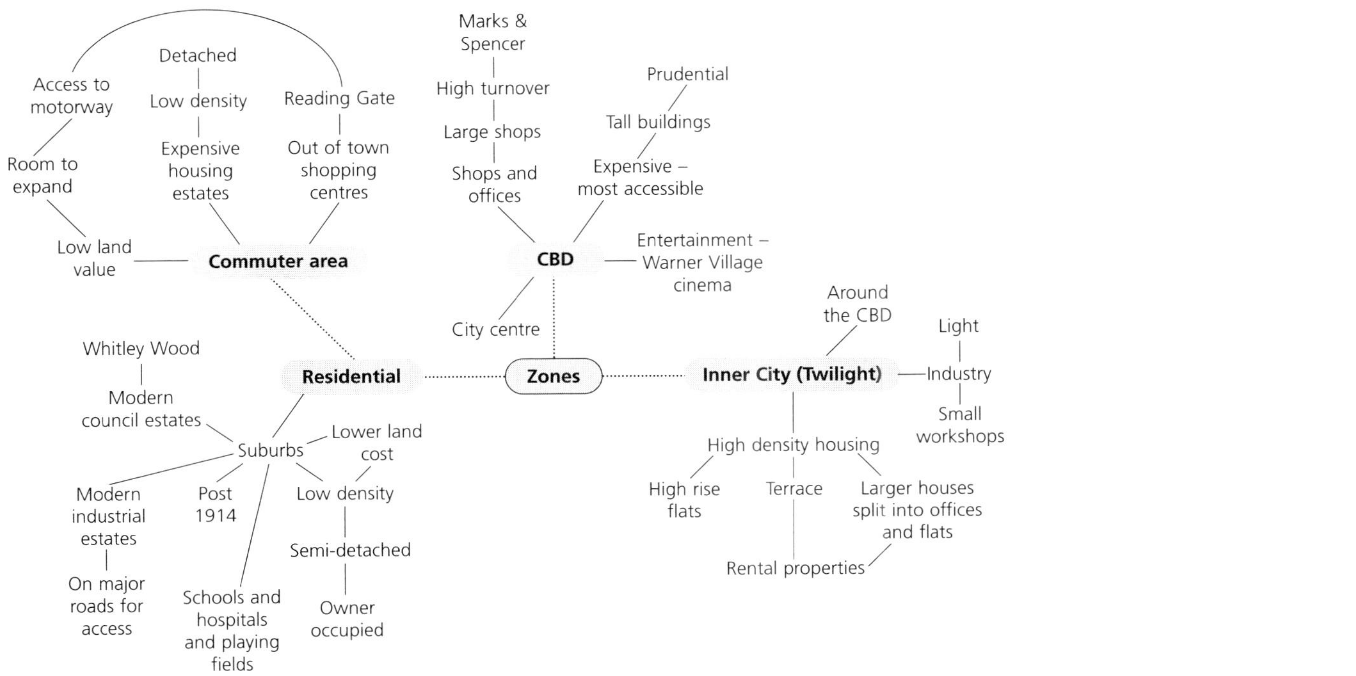

3 The economic world

There are hundreds of different jobs which can be put into three categories:

- **Primary industry** is the extraction of raw materials from the ground or the sea. It includes farming, fishing, forestry and mining
- **Secondary industry** is the making or manufacturing of something using the raw materials from primary industry
- **Tertiary industry** does not produce anything. It often involves providing a service for people. Teachers, solicitors, sales assistants and cleaners are all tertiary occupations

The employment structure of a country is the % of people in each type/category of industry

You don't need to learn these countries but just know what is typical for MEDCs and LEDCs.

LEDCs and MEDCs have different employment structures
Look at the divided bar graphs below.

Key
Primary industry
Secondary industry
Tertiary industry

Mali is typical of an LEDC

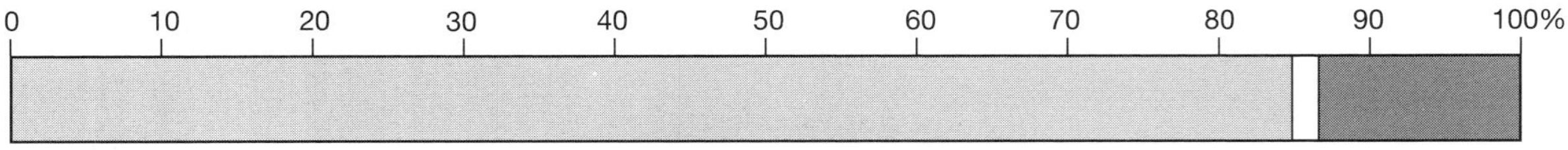

- 85% in primary industry.
- most of the population are farmers.
- very small % in secondary because there is no capital available to develop manufacturing industries.
- small % in tertiary providing services for the population.

Taiwan is a newly industrialised country

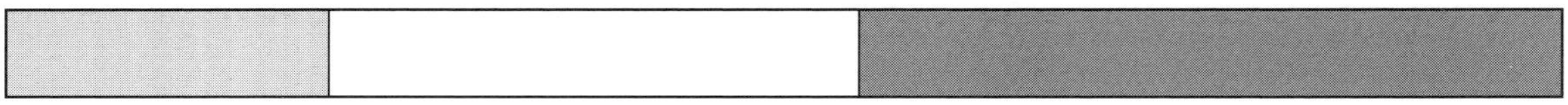

- a fairly large number are employed in primary industry, mainly as farmers, because of lack of mechanisation.
- tertiary industry sector is large and growing.
- 35% of the population is in the secondary sector because many transnational companies have set up in Taiwan because of cheap labour.

Germany is typical of an MEDC

- very few are employed in primary jobs because mechanisation has made a large workforce unnecessary and many raw materials have run out. People also prefer jobs in tertiary industry which have better career prospects.
- secondary industry is at approximately 30%. This has declined with the increase in NICs due to labour being cheaper there.
- approximately 67% of the population is employed in tertiary industry. This means that there is a large service sector. There has also been a large increase in computer technology which comes into this sector.

You need to know how the sectors have changed in different countries and why they have changed

Read the annotations again, they also explain how the sectors have changed over time

Farming characteristics and change

The following are definitions that you need to learn:

Intensive farming – an intensive farm is one that has high levels of labour and / or capital input for a relatively small area of land. The output per hectare is high.
Extensive farming – an extensive farm is one that has low levels of labour and / or capital input for a relatively large area of land. The output per hectare is low.
Commercial farming is when farmers grow crops or rear animals to sell at market to make a profit.
Subsistence farming – farmers grow enough food to feed themselves and their families. There may be a little left over for sale at market.
Arable farms only grow crops.
Pastoral farms rear animals. The only crops grown are fodder crops.
Organic farming – farming that is environmentally friendly. Farmers do not use any chemicals and do not feed their animals with anything that has been grown using chemicals.

The factors that affect a farmer's decision can be physical or human.

The mind map below contains brief notes on the factors

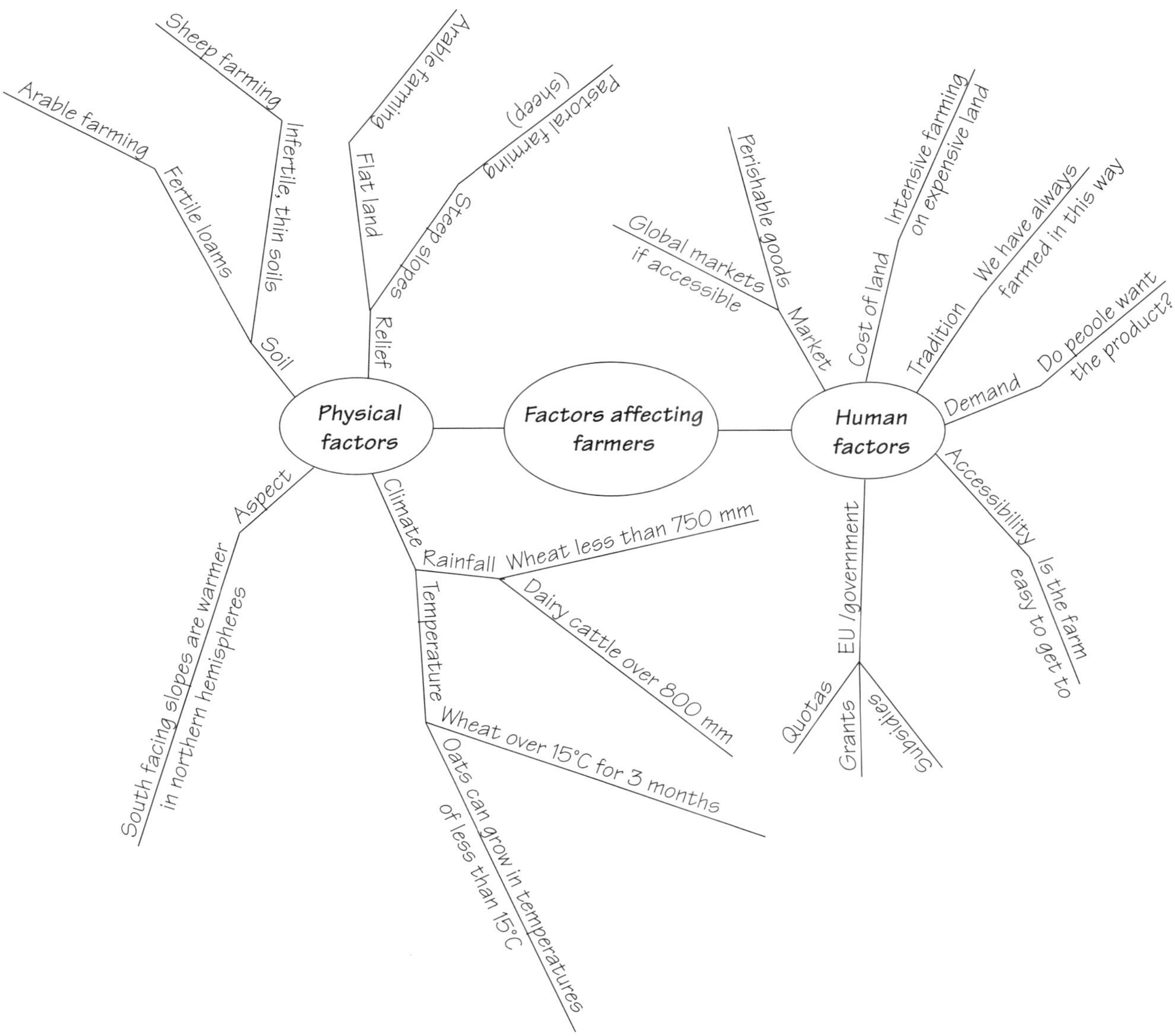

Intensive wet rice farming

You must learn two farming case studies: wet rice faming and a farm in the EU

Case Study

Philippines – north of Indonesia, south east of China

Subsistence farmers, many do not own their own land.
Average farm size 1.5–2 hectares.
Since 1970s the government has been trying to introduce high yielding varieties which require fertilisers – the Green Revolution. Poor farmers could not afford this so only richer farmers benefited. More recent research has been crossing more traditional varieties with faster growing ones with some success.

Two different farms need to be learnt for intensive wet rice farming.

The farms are located on Negros Occidental on the Philippines. The climate has temperatures of 25°C and an average rainfall of 1800mm, 80% of which falls between June and December.

Farm 1 Maximo Casiendo	Characteristic	Farm 2 The Flora Community
2.6 hectare.	**Farm size**	86 hectares.
One family (9 people) mainly subsistence. Some left for sale.	**Farm organisation**	76 workers mainly subsistence. Some left for sale. Organic farming.
Clay loams.	**Soil**	Sandy loams.
Flat, fertile land. 70m above sea level.	**Relief**	Flat, river floodplain.
7–8 bags of nitrogen, 3 bags of complete **NPK** per hectare, 2 bags of **PK** per hectare. Pesticides 2 litres per hectare.	**Chemicals**	Organic farming – weeds controlled by crop rotation, insects are controlled by biological strategies such as preditors. Crops are dried by solar or biomass energy.
Rice, maize, some vegetables and cassava. 2 crops a year.	**Outputs**	Sugar cane, high value vegetables for sale. Rice for the community.
100 cavans per hectare.	**Yield**	100 cavans per hectare.

The Common Agricultural Policy for EU countries

Then	Now
Subsidies – money given to farmers per head of livestock. Set-aside – voluntary when first introduced in 1988. Quotas – on the amount of milk farmers were allowed to produce. Grants – to farmers in environmentally sensitive areas. Guaranteed prices – farmers are guaranteed a certain price for their crops.	Set-aside – in 2000 10% of farmers land must be set-aside if they are to receive arable area payments. Arable area payments – made per hectare to growers of cereal crops. Milk quota – a restriction on the amount of milk that a farmer may produce. Quota on sheep and beef. Woodland grant – grants are available to farmers who wish to plant woods on their land. Environmentally sensitive area – farmers in 22 areas of the country can sign a management agreement which gives them money in return for protecting the environment. Hill farm allowance scheme – paid to farmers who farm land within a moorland boundary.

Changes affecting farming systems in the EU

British farming is experiencing many problems and many farmers are going bankrupt, others are using their land and buildings for other purposes to make a living. This is known as **diversification**.

Case Study

Facts about Home Farm

Owner – Mr Redfern.
Location – Hampton in Arden, SE of the Birmingham conurbation.
Climate – 16°C in July; 3°C January, 700mm rainfall.
Relief – gently undulating land.
Markets – Banbury 60 km.
Animals – 300 sheep and a herd of young bulls being fattened for meat.
Arable crops – mainly wheat and barley.

Problems experienced by Mr Redfern

Dairy herd sold due to poor return.
Subsidies have been decreased leading to a decrease in income of £40 000 between 1998–1999, there was a further decrease in 2000.
Lack of labour – so the whole family help on the farm.

Diversification

This has been occurring over the last 5 years as subsidies have been decreased by the EU.

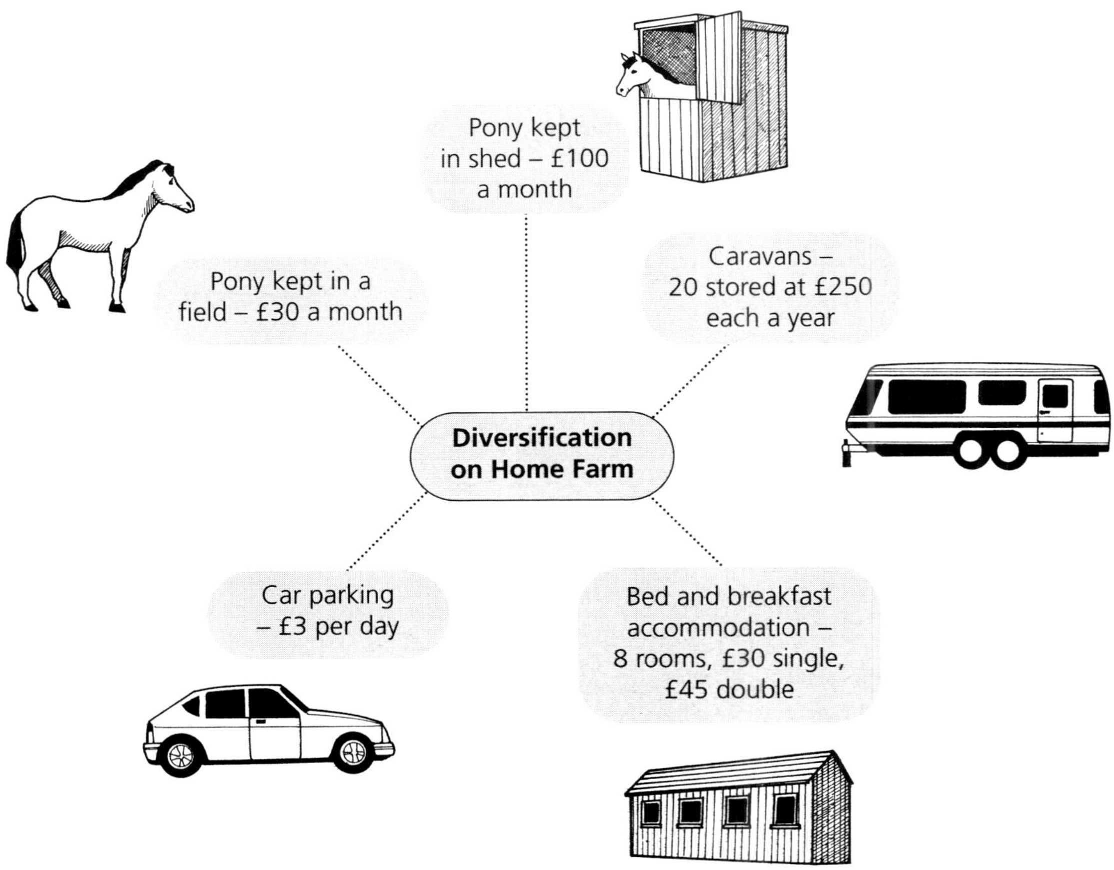

Factors affecting the location of industry in MEDCs

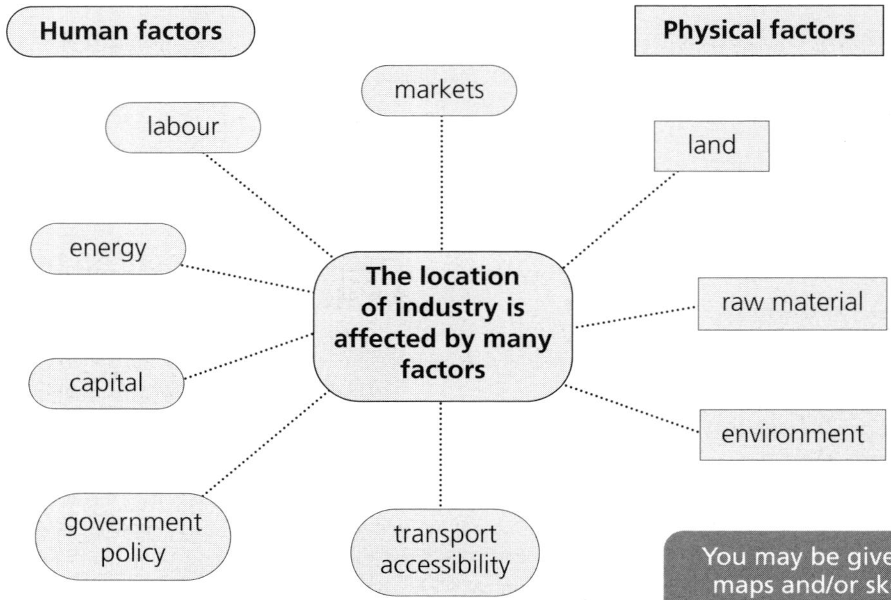

You may be given OS maps and/or sketch maps/photographs to investigate locational factors

High tech industry case study – computer development in Bracknell

The following factors were important in influencing the location of the high tech industry in Bracknell.

The map below has all the information that you need.

Bracknell has plenty of flat land available for development

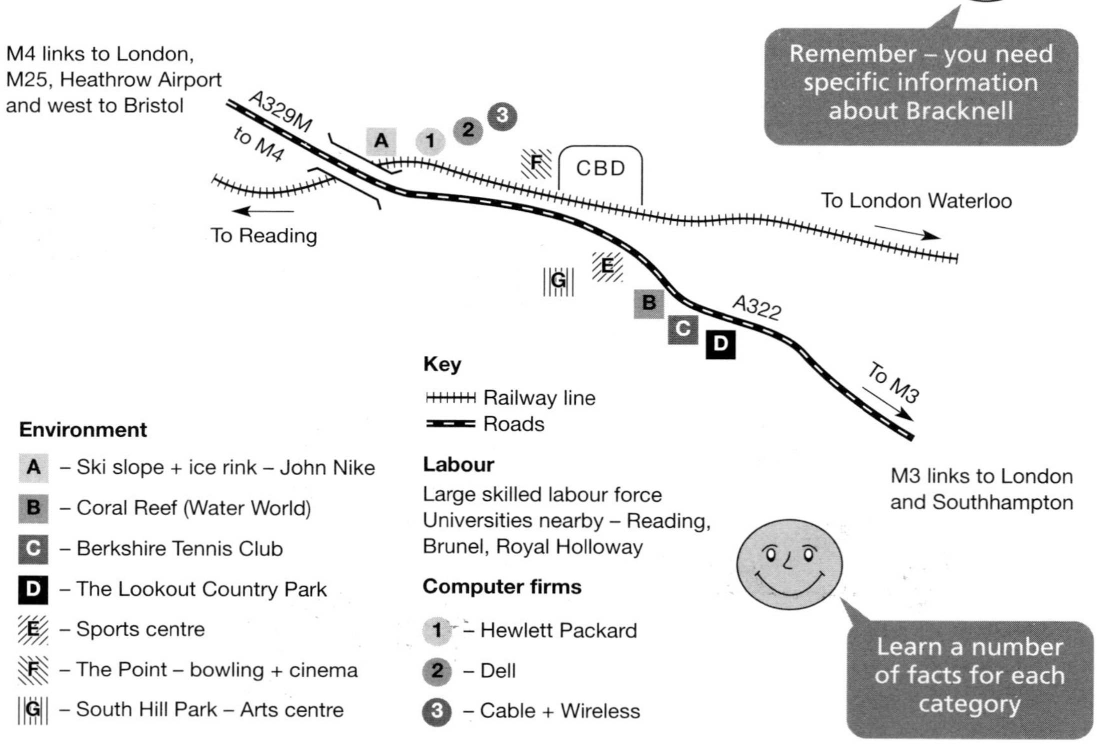

Industry in LEDCs

You must know the difference between the informal and formal sectors in LEDCs.

Formal	Informal
• Fixed hours	• Labour intensive
• Fixed wage	• Small scale
• Pay taxes	• Few skills required
• Requires certain skills	• No fixed hours or wages

The exam question would say 'for a TNC you have studied why did it locate in a particular country?'

Your case study is Fiat and what attracted Fiat to Brazil.

Case Study

Fiat has bases in 64 countries

Its base in Brazil is at Betim. It opened in 1976, employs 10,000 workers, and produces 130,000 cars a year. In 1990 Fiat invested another $1.5m in the plant to make Palio and Tipo cars. It now employs 12,000 and produces 650,000 cars a year.

Why did Fiat locate in Brazil?

The Brazilian government wanted a car industry and it offered incentives to large TNCs

↓

Betim is in the state of Minas Gerais. Fiat located there because the state gave them 50% of the start up costs

↓

Fiat could supply Brazil's increasing market for cars. 1970 – 255,000; 2000 – 2,800,000 cars produced and sold

↓

Cheap labour: $7 an hour. In UK workers get $18 an hour

↓

Strikes are very unlikely so no production time will be lost if Fiat locates in Brazil

↓

Fiat has recently expanded – why?

↓

The state of Minas Gerais gave Fiat $135m towards building a new factory – over half the cost

↓

Fiat are expecting a large increase in demand. If they are in the country they will save money on transport costs and import duties

Learn the facts that go with the reasons

4 The natural world

- An air mass is a very large volume of air with similar temperature and humidity characteristics
- The weather that an air mass brings to the UK is influenced by:
  a, where the air mass comes from (warm or cold area)
  b, what it passes over (land or sea)
- The weather in the UK is changeable because several different air masses affect the country

Which air masses affect the UK?

The diagram shows the five air masses that you need to know. You will need to know the typical weather that each air mass brings to the UK

Name of air mass	Arctic maritime
Source region	Arctic ocean
Characteristics	Cold and wet, rare in summer
Weather conditions	Freezing conditions in winter, heavy snow and strong winds

Name of air mass	Polar maritime
Source region	Canada and Greenland
Characteristics	Wet and cold
Weather conditions	Cool conditions throughout the year. Clouds and showers along the west coast in winter. Thunderstorms common in summer

Name of air mass	Polar continental
Source region	Scandinavia and Russia
Characteristics	Cold in winter, hot in summer
Weather conditions	Brings snow to the east coast in winter, dry weather in summer

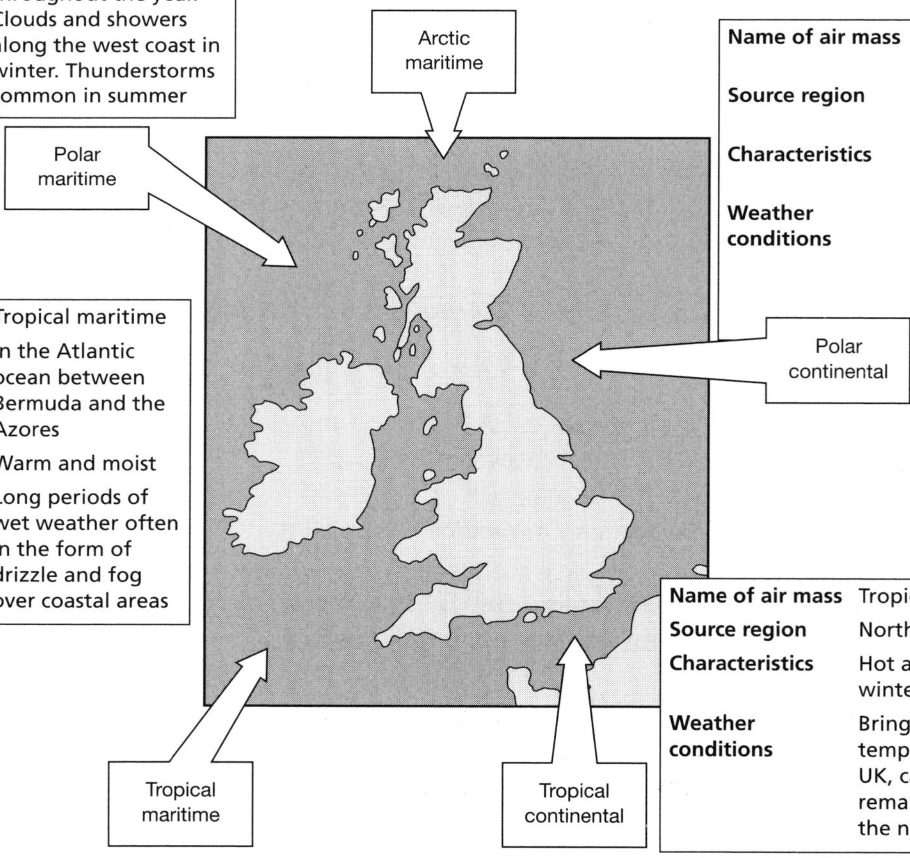

Name of air mass	Tropical maritime
Source region	In the Atlantic ocean between Bermuda and the Azores
Characteristics	Warm and moist
Weather conditions	Long periods of wet weather often in the form of drizzle and fog over coastal areas

Name of air mass	Tropical continental
Source region	North Africa
Characteristics	Hot and dry, rare in winter
Weather conditions	Brings the highest temperatures to the UK, can reach 30°C and remains high during the night 15–20°C

Rainfall in the UK

Precipitation is the name given to all types of moisture that fall from the sky.

Rain is formed when:

- Moist air is forced to rise
- Water vapour in the air cools and condenses and forms clouds
- Droplets within the clouds coalesce (join together) and when they are heavy enough they fall due to gravity

There are three main types of rainfall

1. Relief rainfall

Warm moist air, which has blown across the Atlantic, cools and condenses as it is forced to rise up the hills on the west of the UK. Clouds form and rain falls over the hills. As the air descends it warms, becomes drier and clouds disappear. This creates a rain shadow on the eastern side of the UK.

2. Frontal rainfall

When two air masses meet, warm moist air rises over colder air. This warm air is less dense than the cold air. It cools and condenses, forming clouds, which release rain. This type of rain is common throughout the UK and at all times of the year.

Think of the steam from a boiling kettle condensing as it hits a cold surface like a window. The droplets join together and run down the window

3. Convectional rainfall

When it is hot, air warms and rises. As it rises the air cools and water vapour in the air condenses to form clouds, which may result in heavy rainstorms with thunder and lightning. This type of rainfall is most common in the summer in southeast England.

Features of an anticyclone?

- Anticyclones are areas of high pressure
- They give mainly dry weather because air is sinking and warming up, therefore there will be few clouds making rain less likely
- Winds are very light, they blow clockwise around the anticyclone
- They are large areas often affecting the whole of the UK
- They may last several days

Note that the weather station at X has no cloud cover, is calm and the temperature is 25°C

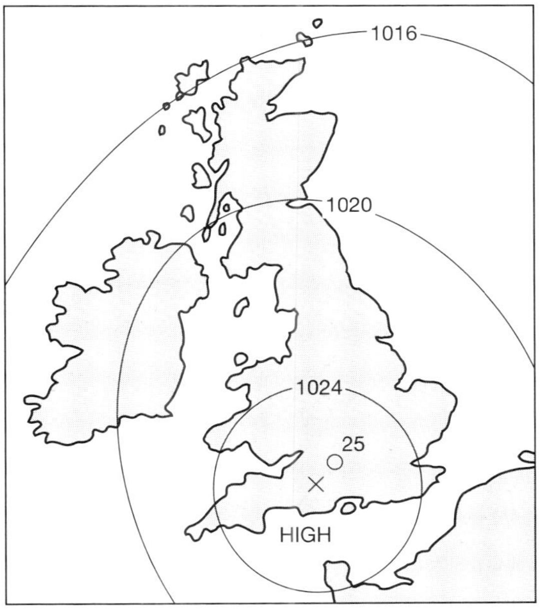

There are differences in anticyclonic weather in the UK between summer and winter.

	Summer	Winter
Temperature	Higher than average Heatwaves possible	Lower than average Big freeze possible
Precipitation	Dew and mist at night Thunderstorms often signal the end of the anticyclone	Frost and fog at night

What weather is associated with depressions?

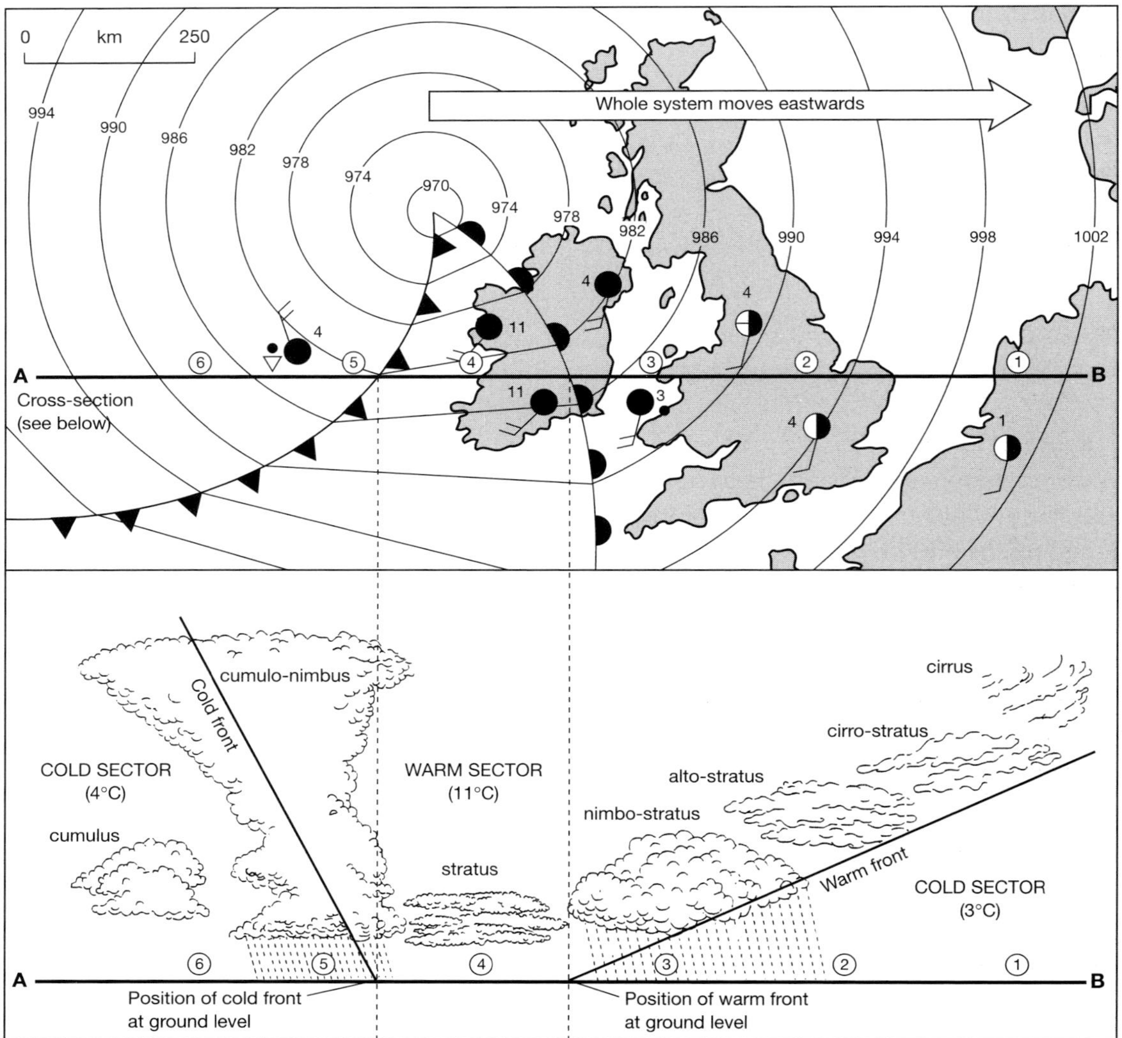

	6	5	4	3	2	1
	After cold front	As cold front passes	Warm sector	As warm front passes over	As warm front approaches	Well before warm front
Cloud type	Fair weather cumulus	Towering cumulo-nimbus	Dull, low, flat stratus	Dense nimbo-stratus	Lower, thicker alto-stratus	High altitude cirrus and alto-stratus
Rainfall	Showery	Heavy	Drizzle or absent	Strong showers	Drizzle	None
Temp (°C)	4°C	4°C	11°C	3°C	4°C	1°C
Wind	NW	NW	SW	S	S	S
Air pressure	Rising	Rising	Steady	Falling	1002-998 falling	High (1002), falling

If there is a weather map in the exam you will not need to remember all the symbols, as a key will be provided. You will however be expected to recognise warm, cold and occluded fronts

The diagram explains:

- What a depression looks like on a synoptic chart (weather map)
- A cross section of the depression
- The weather conditions in the different sectors

Climate types

Climate is the average pattern of weather for a year.

The monthly changes can be plotted on a climate graph.

You may be asked to complete a climate graph

You may be asked to describe the distribution of rainfall and/or temperature over a year. The examiner will be looking for general comments and more specific detail. If data is asked for you will lose marks if you don't include it. It is a good idea to include it even if it is not asked for

What are the characteristics of the west European maritime climate?

This type of climate is experienced between 40°–60° north of the equator. This climate area spreads from the UK at the extreme west through the lowland areas of West Europe including the coastal fringe of Scandinavia to upland areas such as the Alps which act as a natural barrier to the eastward spread.

The characteristics can be best revised by reference to an example.

Below is a climate graph for Oxford, around it are comments which sum up the characteristics of the west European maritime climate. **LEARN THEM!**

A climate graph for Oxford

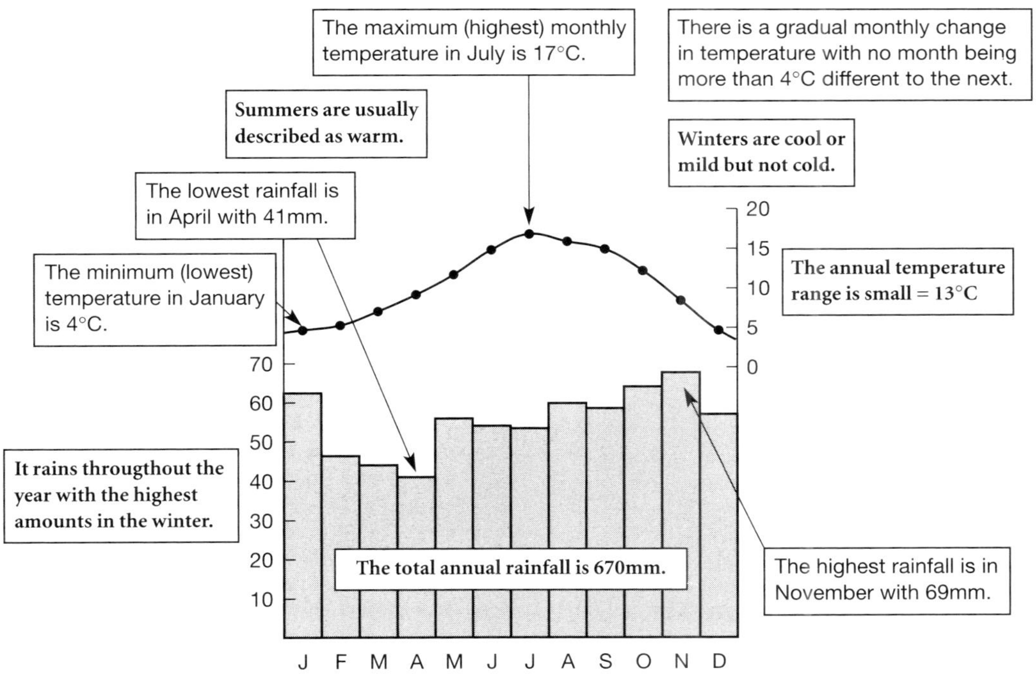

What are the characteristics of east European continental interior climate?

The total annual rainfall is the total of the monthly rainfalls

Areas experiencing this type of climate occur inland between 35° and 60° north of the equator.

Again you are best to relate the information to an example. Below is the climate graph for Kiev. It is a typical east European continental interior climate.

A climate graph for Kiev

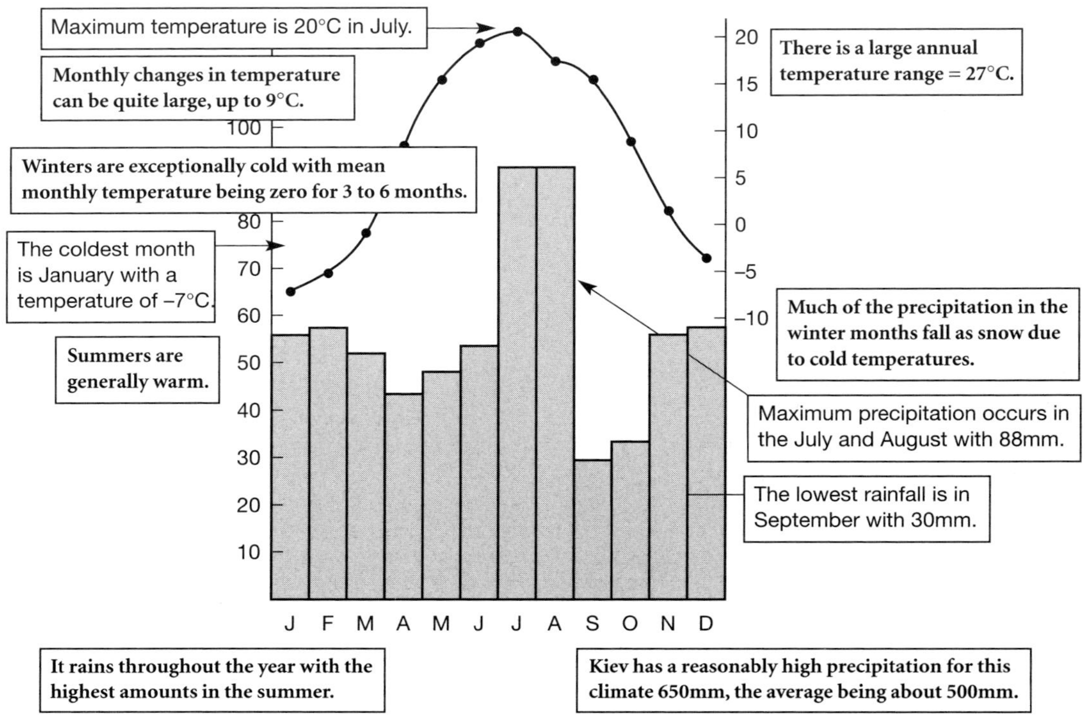

The annual temperature range is the difference between the highest and lowest temperatures

Learn the labels around it so that you can describe the climate type

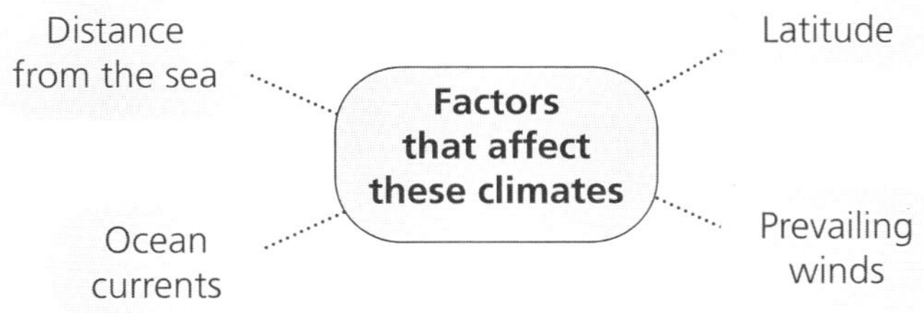

Distance from the sea

Places that are close to the sea have their temperatures moderated by its effect. This means that they are warmer in winter and cooler in summer than places further inland.

But why?

Autumn and winter influence

Imagine a football match. The players on the field will lose temperature at different speeds. The goalie will lose heat quickly because he is still (if his defenders are any good!). The rest of the players are moving around therefore they will lose heat more slowly because their movement generates heat and energy.

Which is the goalie the land or the sea?

Now let's come back to reality. The goalie is still like the land and will lose heat quickly. Therefore places that are inland cool more quickly in the winter than places by the coast which are warmed by winds blowing off the sea.

Spring and summer influence

The sun has now come out. The goalie is standing still; the players are moving. Which is going to warm up the quickest?

The goalie will but only on his front, just like the land surface. The players who are moving, like the sea, will warm up more slowly but all of their bodies will warm, not just one side like the goalie. Therefore in the spring and summer coastal areas are cooled by winds that blow off the sea because the sea is cooler than the land as it takes longer to warm up.

Latitude

Places close to the equator are warmer than places near the poles. This, as you know, is a purely descriptive comment on temperature. We now need to think why?

- The sun's energy is more concentrated because it has less area to warm up
- The sun is at a more direct angle to the equator than the poles and therefore has to pass through less atmosphere. Less heat is lost
- The length of day is also important. During winter in the UK the days are short with few sunlight hours therefore temperatures are bound to be lower

Ocean currents

- Ocean currents are either warm or cold depending on where they come from
- Cold currents move towards the equator
- Warm currents move away from the equator
- The warm North Atlantic Drift current raises temperatures in coastal western Europe e.g. France and the UK

Prevailing winds

- The prevailing winds affecting the UK are from the south and west and are therefore warm and wet having travelled across the Atlantic
- The prevailing winds affecting continental Europe are cold dry easterlies

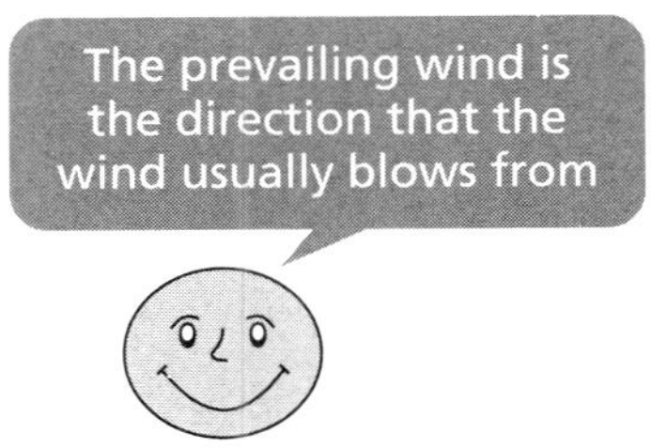

Describe the global distribution of tropical rainforests, temperate deciduous and coniferous forests

What does this mean?
It is asking you where in the world you would find the forest types. There could be a map to help you or just a question

Phew! No need to explain this one. It's a doddle!

How do you describe a distribution?
Follow these easy steps. This description is of the tropical rainforest areas.

- Start with a general statement. **Tropical rainforests are found between the Tropic of Cancer and the Tropic of Capricorn**
- Then give more specific detail such as countries, continents and directions. **The largest area is in South America mostly in Brazil. They can also be found in West Africa in Cameroon and Zaire. There are areas in SE Asia and Indonesia. Other rainforest areas can be found in Queensland in North East Australia**

Temperate deciduous forests

- Found between 30° and 60° north of the equator
- Found mainly in the east of the USA. They are found in Europe, for example Germany, southern Sweden and Norway. They are also found in eastern Asia in China and Japan
- Not in the southern hemisphere due to lack of land at the correct latitudes

Temperate coniferous (taiga) forest

- Found between 50°N and the Arctic circle
- Found in a belt stretching across Canada and northern Europe through central Norway, Sweden, Finland and across Russia
- Not in the southern hemisphere because there is little land at the correct latitudes

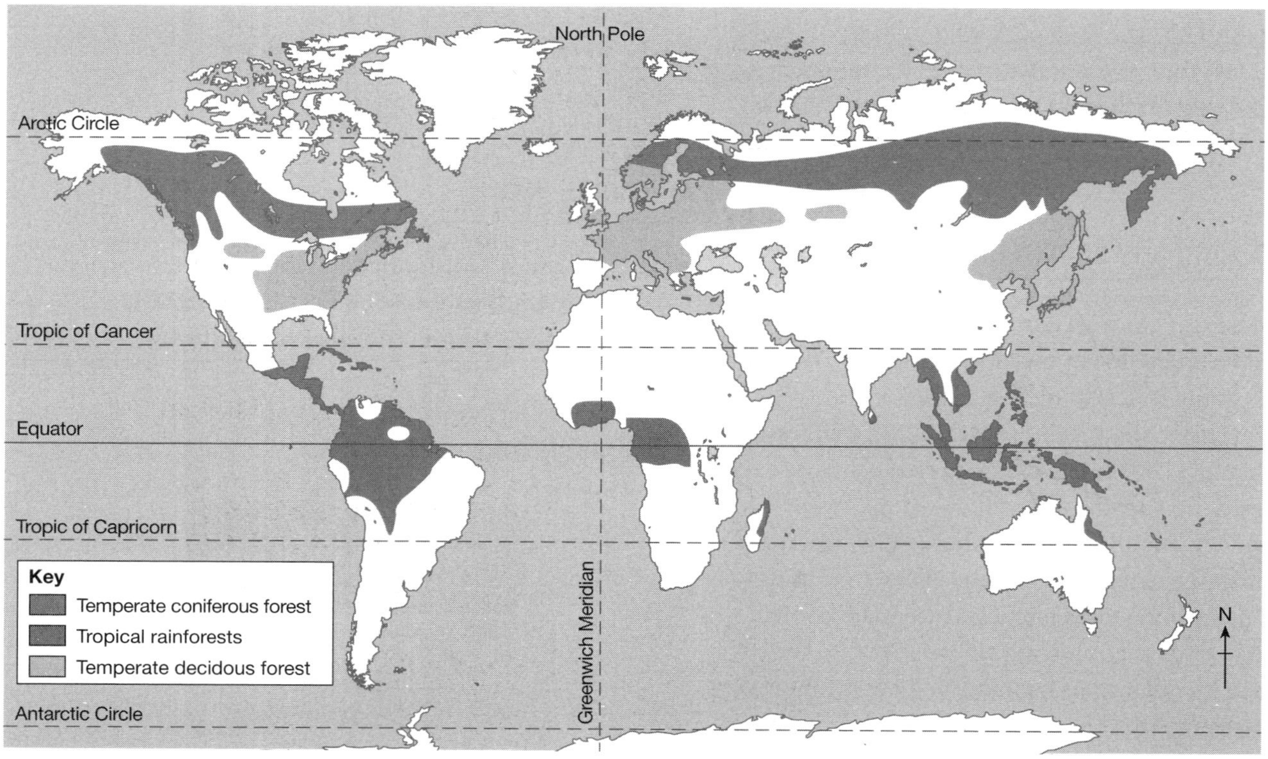

How have coniferous trees (taiga) adapted to the natural environment?

Look for the **adaptation** and *explanation* around the trees below.

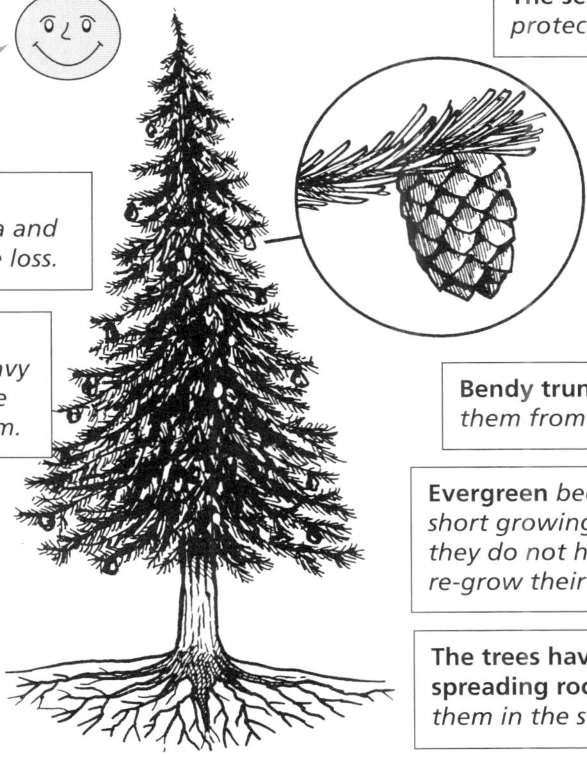

The seeds *are protected in cones.*

Leaves are needle shaped *therefore small surface area and waxy. This reduces moisture loss.*

The branches slope downwards *to stop the heavy snow from collecting on the branches and breaking them.*

The trees are conical shaped *this gives them stability in high winds, common in these areas.*

Bendy trunks *protect them from high winds.*

Evergreen *because of short growing season they do not have time to re-grow their leaves.*

The trees have wide spreading roots *to hold them in the shallow soil.*

You could be asked to explain the annotations. Simply stating them will not earn many marks

The questions could ask for annotations to be put onto a tree as shown

How do trees in the tropical rainforest adapt to the natural environment?

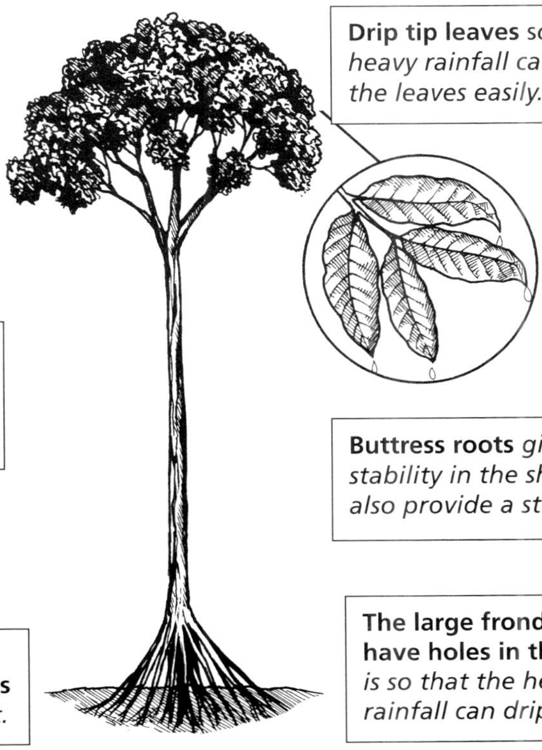

Trees grow very tall and straight *this is to reach the light as quickly as possible.*

Drip tip leaves *so that the heavy rainfall can run off the leaves easily.*

They have few branches on their trunks. *This is due to the lack of light in the lower forest.*

Buttress roots *give the trees stability in the shallow soil. They also provide a store of nutrients.*

Plants in the lower forest have large frond-like leaves *because of the lack of light.*

The large frond leaves have holes in them. *This is so that the heavy rainfall can drip through.*

A study of the tropical rainforest to show the ideas of conservation, exploitation and sustainable development – Papua New Guinea

Conservation – to manage the environment in such a way that it will be protected from change.
Sustainable development – the development of an area using techniques and approaches that will help to preserve the environment for the future.
Exploitation – when the environment is used in such a way that it is destroyed and will be of no use to future generations.

Remember learn the factual detail

In this case study you could be asked questions about conservation, sustainable development and exploitation

Exploitation

Papua New Guinea in South-East Asia. The land is covered in tropical rainforest with 9000 species of plants, 250 species of mammals and 7000 species of birds.

- The government of Papua New Guinea sells areas of the rainforest to companies with little thought for the environment or local people
- Aitape – 1996 timber authority given to Pia-Damansara to clear fell 5000 hectares of forest for hardwood
- Sissano lagoon – 1996 large area of forest which belonged to the Sissano community was cleared to grow palm oil. They also lost prawn fishing grounds and there is concern about the coral reef being affected

Conservation

- Crater mountain wildlife management area set up in early 1990s covers 2600 sq kms. Within this area there are 220 species of birds and 85 mammal species. A small community in the area is threatened by tourism and logging. By declaring it a wildlife management area it will be preserved. Guides and eco-lodges are provided for tourists.

You could also be asked about conflicts, don't forget to write about groups of people.

Sustainable development by the government

- 2000 prime minister of PNG said that no new logging licences would be granted
- This was not implemented. Illegal logging continued
- World bank threatened to withdraw funds if the government did not control the felling of the forest
- New approach: controlled logging as part of management strategies is allowed
- Processing plants will be built
- Community forestry which is sustainable will be encouraged

Sustainable development by NGOs

- Kikori basin 1993–2000. Chevron drill for oil in this area. Also working with the local community on sustainable activities such as nature-based tourism, fish and butterfly farming, and developing cottage industries in the area
- Collingwood Bay project 1996–2003. This area already has tourism so ecotourism is being encouraged. Cottage industries are also being developed such as Tapa cloth production. Greenpeace has helped with this, developing markets in the USA. Nut harvesting will be encouraged. Maisin community was given grants to help out with projects.

5 Managing the environment

This chapter includes coastal defences, river defences and sustainable management of fragile environments.

You must know the most common forms of coastal and river defence used in the UK. And be able to describe them

Learn four case studies for this part of the course

Know the difference between hard and soft engineering techniques

The causes of coastal recession

OR why do some cliffs fall down faster than others?

- Rock type – harder rocks are eroded more slowly e.g. granite
- Rock angle to the coast – parallel or at right angles
  Think of headlands and bays!
- Weathering and mass movement
- Erosional processes – hydraulic action, corrasion and corrosion
- Fetch – the distance the wind blows over open water
- The longer the fetch the stronger the wave
- Beach shape – long gently sloping beaches absorb wave energy

Hurrah!
No need to learn these because you already know them from chapter 1.
You hope!

The causes of flooding

These can be split into physical and human.

Physical

- Large amounts of continual rainfall, causes saturated ground and greater run-off
- Cloudbursts, rain droplets too large and fall too fast to sink into the ground, so greater run-off
- Rapid snow melt, greater run-off
- Silted up river channels can hold less water
- Impermeable rock surface, greater run-off
- Steep valley sides, faster run-off

Human

- Removal of vegetation, less interception, greater run-off.
- Towns built on floodplain, tarmac is impermeable, greater run-off
- Global warming, melting of polar ice caps, rise in sea levels flooding of lowland river valleys
- Dams may burst

Remember!
In most questions you will be asked for case study material on the causes of river flooding

Other types of coastal defence which will be credited but are not dealt with on the next page

- offshore reefs
- seabees
- stabilising dunes
- rock armour (rip rap)
- revetments

Hard engineering techniques

Soft engineering techniques

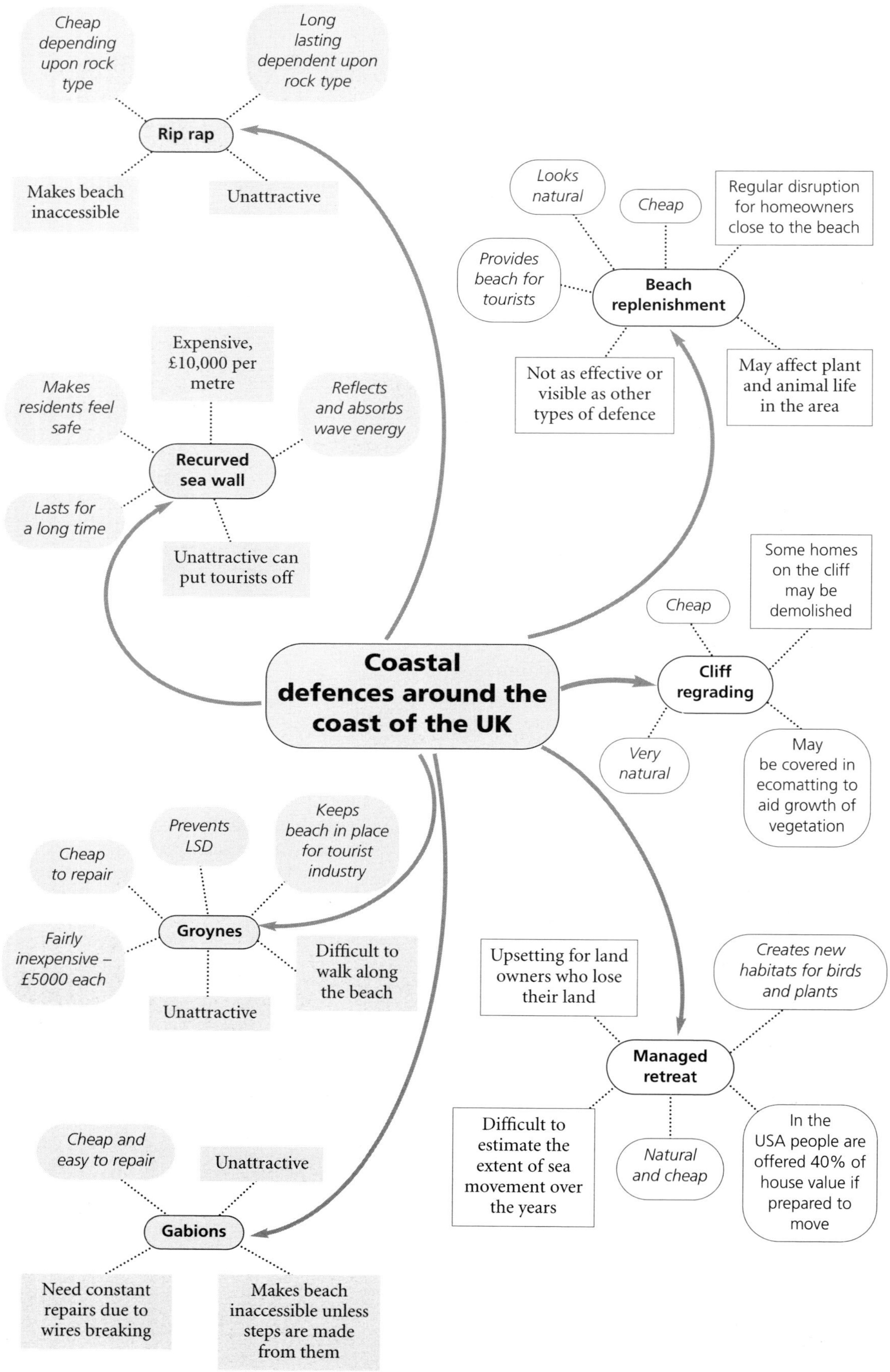

Case Study

The causes, effects and management of cliff recession – Walton-on-the-Naze

Causes

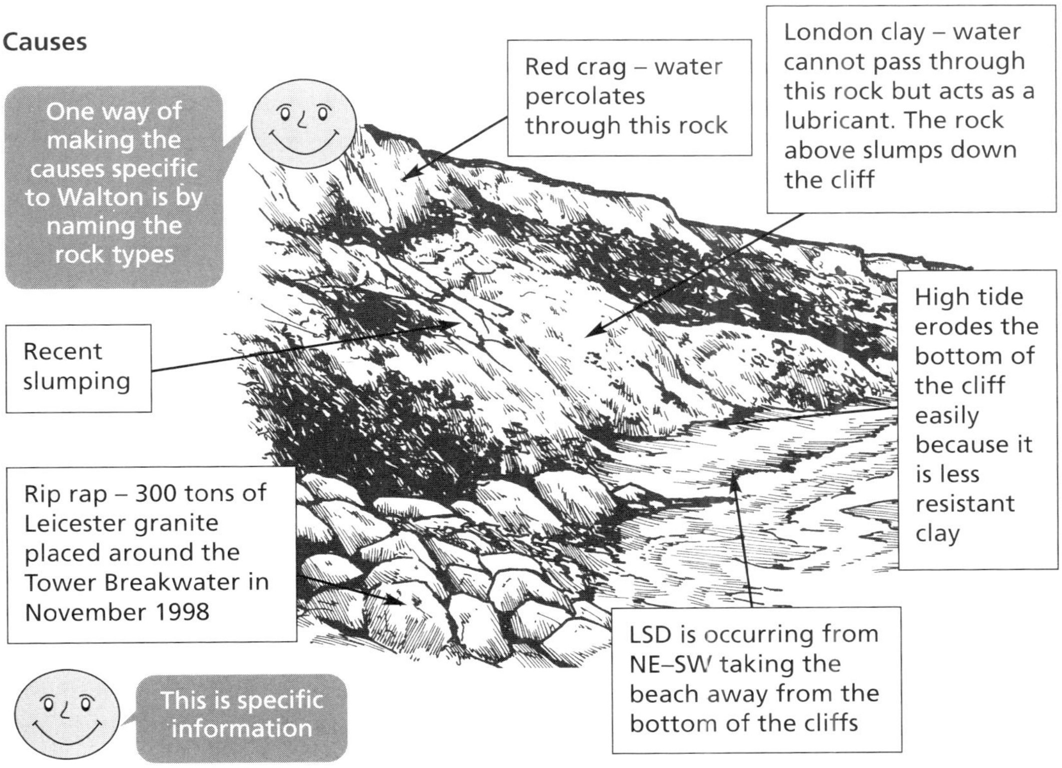

Effects – what has happened

- Pill boxes on the beach were on the cliff top in 1945; now 35 metres from the cliff
- Students results show that the cliff has retreated 9.6 metres in 6 years

Management

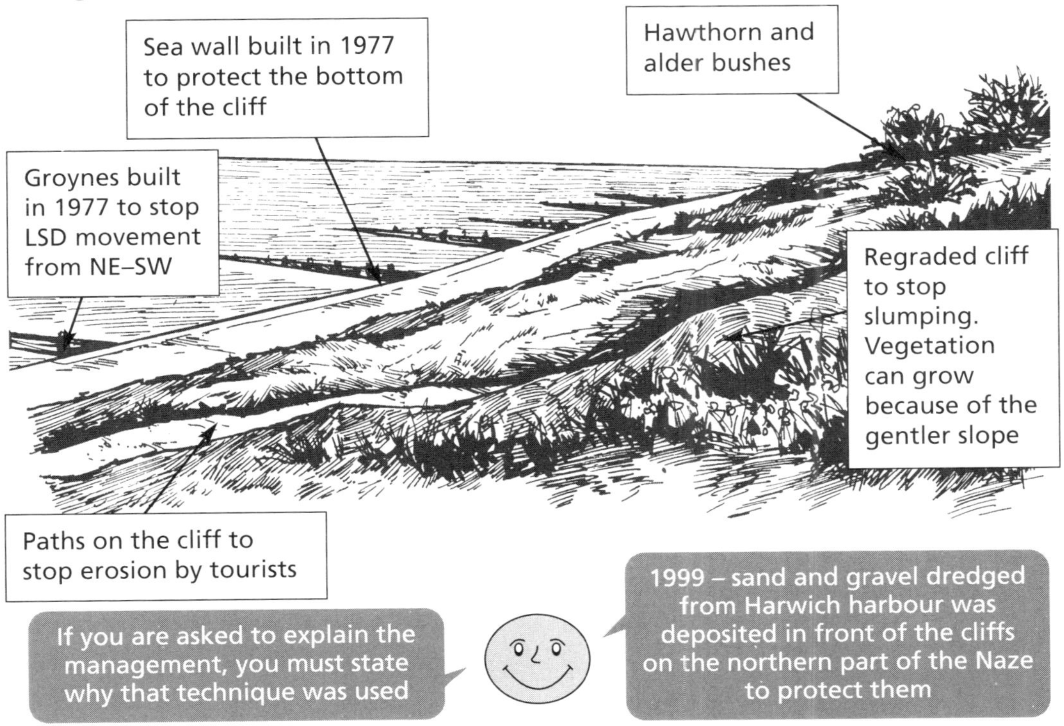

If you are asked to explain the management, you must state why that technique was used

1999 – sand and gravel dredged from Harwich harbour was deposited in front of the cliffs on the northern part of the Naze to protect them

Remember you could be asked
➩ What they look like?
➩ Their advantages or
➩ Their disadvantages
Be prepared!

River defence techniques – you need to know the most common defence techniques that are used on rivers in the UK. Check out the diagram below.

The hard options

Do you know the difference between soft and hard engineering techniques? Check out the first page of this chapter

Dams

- These are built across the river usually in the upper stage
- They allow people to control the flow of the water and make floods less likely
- They are expensive to build
- They can only be built in certain areas where rock type and river basin shape are favourable

Flood relief channels

- These are built to take water out of the river at times of flood
- This is a channel which water can be allowed into via sluice gates if a flood is likely. The channel meets with the river further downstream usually the other side of a built up area and allows the water back into the river after the peak flow has passed

The soft options

Afforestation

- Trees are planted in the drainage basin
- The trees intercept rain water before it can reach the river channel
- This means that there is less chance of flooding downstream
- This can only be done if there is room in the drainage basin
- Generally this is a fairly cheap option

Land use zoning, washlands and distance embankments

- The land use around the river is strictly controlled
- The land right next to the river is used for less important uses. If there is a flood this land is allowed to be covered in water and uses up some of the rivers extra capacity
- Distance embankments are then built to stop the river flooding too far onto the floodplain
- If the land is available this is a fairly cheap option
- There can be opposition from local farmers who may periodically lose their land to flooding
- It works with the river maintaining the natural meanders and river channel

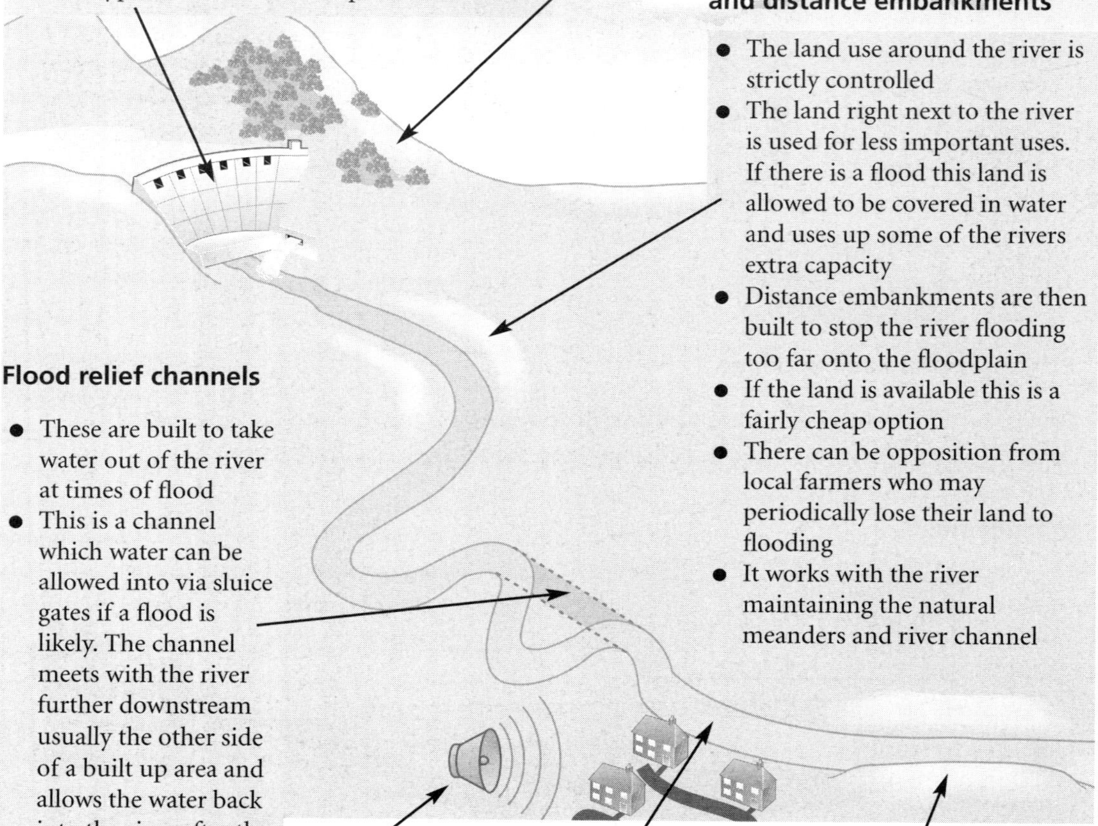

Flood warning systems

- In the UK the environment agency warns householders if a flood is likely to occur in their area
- They don't really solve the problem of flooding however people can take precautions if they know that a flood is likely to occur

Channelisation

- This is the deepening and widening of river channels
- Makes residents feel safe
- Expensive and needs regular maintenance
- Can take up land next to the river
- The machinery needed is noisy and disruptive to local people

Building embankments/levees

- These are walls or earth banks built either side of the river
- They make people feel safe
- They are expensive to build and can take up a lot of room next to the river

Case Study

The causes, effects and management of river flooding – the Mississippi Flood 1993

It is best to use a specific example and learn the facts. There are lots of facts on each section

Causes

- January–June 1993. North Dakota, Kansas and Iowa received more than double their normal rainfall
- 4th and 5th July, 160mm rain fell in parts of southern Iowa
- Saturated ground caused greater run-off
- 80% of the original wetlands have been drained since 1940 – for farmland and the growth of towns
- The straightening of the channel has caused the river to flow more quickly. Therefore water reaches the lower valley earlier

Effects

- 32 people were killed, 30,000 evacuated
- 60,000 homes were destroyed or damaged
- $10 billion of damage to property
- 4m hectares of farmland was flooded
- Corn production went down by 10%
- Des Moines lost its water supply for 19 days
- Barge owners who used the rivers to transport goods such as coal lost $1m a day
- Hundreds of miles of roads and bridges were closed

Simple answers such as levees built will score low marks. Levees built along 3,000 km of the River Mississippi is specific and therefore scores high marks

Management before the flood

A* should aim to learn a minimum of 4 causes, 4 effects and 4 types of management

105 dams have been built on the Missouri.

Levees built along 3000 km of the river, some up to 15m high.

1750 km of the river has been straightened. The channel has been lined with concrete.

Illinois Groundwater Consortium are removing land from agricultural use and re-establishing Wetlands close to the river.

Afforestation of the valley sides.

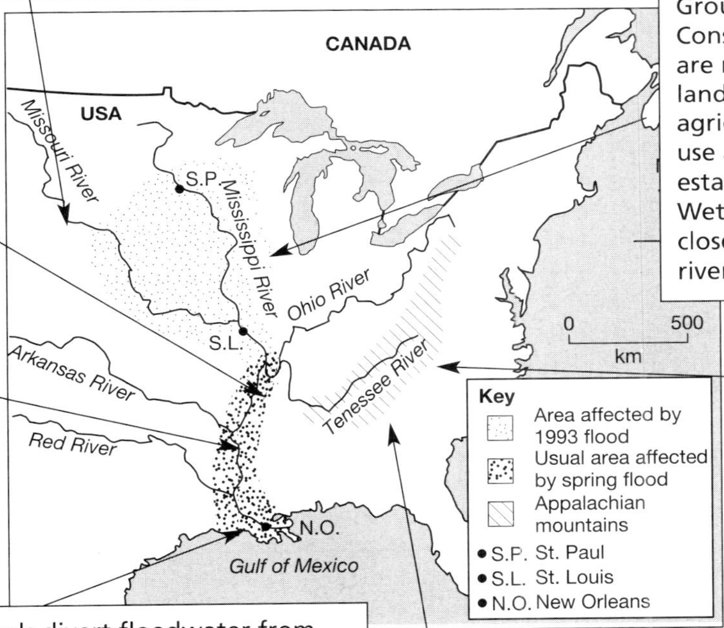

Overflow channels divert floodwater from the Mississippi. Bonne Carre floodway, 50km north of New Orleans takes water straight into Gulf of Mexico via Lake Pontchartrain

Six large dams built on Tennessee river.

For a C grade aim to learn 2 causes, 2 effects and 2 types of management

Management after the 1993 flood

- 13 dredgers were utilised to make the river navigable again for large vessels
- By December all breaks in levees were repaired. 150 breaks over 23 miles
- By the end of the year all levee erosion had been repaired and pumping stations were back in working order

Fragile environments require sustainable management

What do you need to know?

You must learn two case studies of damage to fragile environments.
One must be about the damage caused by agriculture (farming).
The other about damage caused by resource exploitation.
First you need some general information about the damage that farming and resource exploitation can cause.

Damage by farming

There are 5 ways that farming can affect fragile environments.

Fertilisers	Pesticides and herbicides	Hedgerows	Irrigation	Overgrazing and over cultivation
Cause eutrophication. Farmers spray fertiliser on their land. Some of the nitrates are washed into rivers. This causes algae bloom. Bacteria feeding on the algae multiply and use up the oxygen in the river. Fish and other river life die. The algae spreads across the surface of the river. This excludes light from the bottom. Plants cannot photosynthesise and the river dies.	These chemicals are sprayed onto crops to ensure healthy growth. Insects are killed by pesticides this disrupts the food chain. Herbicides kill many species of plants that are native to the UK. This causes a loss of biodiversity.	These are removed to allow more space for crop growing. Hedges act as natural windbreaks. The wind is now stronger so can pick up, dry soil and carry it long distances as there is nothing to stop it. Hedges are home to many animals and insects. Their removal means a loss of habitat to many species.	Water removed to spray crops. Water levels are lower. This causes problems for vegetation on the banks of rivers. The banks also become more susceptible to boat wash.	Land is overgrazed when cattle continually graze the same piece of land until they eat the roots as well as the grass. Therefore the grass is unable to re-grow. Over cultivation is when the same piece of land is continually used to grow crops without the use of fertilisers. This leads to a loss of fertility and a decrease in yields.

Damage by resource exploitation

This is the removal of minerals from the ground. This can cause a number of different problems.

Extraction	Transportation	Use
This causes the destruction of large areas of land and the loss of habitats.	The moving of resources from the place of extraction to the place of use can be dangerous to the environment. There have been many oil tanker disasters over recent years.	Fossil fuels when burnt release carbon dioxide which has been linked to Global Warming.

Case Study

Damage by agriculture (farming) in an MEDC

You will be asked questions on causes, effects and management of the damage

The Norfolk Broads

CAUSES

Agriculture

Irrigation – the average farm requires 8m gallons of water from irrigation a year

EFFECTS: Lowers water levels, banks more likely to be eroded

EFFECT: Vegetation on river banks dies

Eutrophication – nitrates run off into rivers. This causes algae to bloom. Bacteria feeding on algae use up oxygen in the water

EFFECT: Lack of oxygen causes fish to die

EFFECT: **Barton Broad**

- number of fish has declined
- age of fish has declined
- variety of wild fowl have declined

EFFECT: Water plants cover the surface blocking out light

EFFECT: River plants die due to lack of light

Sewage disposal

EFFECTS: Adds to the nutrient in the Broads and rivers. Speeds up eutrophication

Tourism isn't farming. But boat wash is made worse by low water level – caused by irrigation

The wash from tourists speed boats

EFFECT: Erosion of the river banks

Loss of the vegetation on the banks of the river. Fringing reed is also lost on the Broads, e.g. Barton Broad

MANAGEMENT

FARMERS: Nitrate levels are being monitored; farmers encouraged to use less chemical fertiliser

FARMERS: Grants for environmentally friendly farming

- arable land restored to permanent pasture
- arable land next to rivers or Broads to have a 6 metre border of grassland around it
- money given to them for planting hedges, creating ponds

RESTORATION: **Clear Water 2000 project at Barton Broad on the River Ant**

- suction dredging of 300,000 cubic metres of nutrient rich mud
- fringing reed is being replaced
- pleasure island is being rebuilt
- new educational facilities
- biomanipulation of the Broad

TOURISM:

Dinghies encouraged, speed boats restricted use

Speed limits to protect banks

Case Study

For this case study you need to know specific points on the damage, the effects of the damage and the management of the problem

Damage by resource exploitation in an LEDC

Oil extraction in the Amazon Rainforest, the Oriente region of Ecuador

THE MAIN CAUSE	THE SPECIFIC DAMAGE CAUSED TO THE AREA	THE EFFECTS	MANAGEMENT
Oil extraction by Texaco since 1964	• 16.8m gallons of crude oil have been spilt in the forest • roads have been built through the forest • pipelines have been put through • 20 billion gallons of waste water containing toxic hydrocarbons have been put into the Amazon's waterways • 1m hectares of rainforest have been destroyed	• disruption to food chains as toxins concentrate in top carnivores • thousands of settlers (the colonos) have moved into the area, encouraged by the Living Frontiers policy. Creating farms in the forest • loss of rainforest vegetation which has species which are not found anywhere else in the world. Many of these species are important medicines for example quinine which can be used to relieve pain • the concentration of hydrocarbons in the water is well above safety levels. There has been an increase in cancer and miscarriages amongst indigenous peoples such as the Huaorani tribe • loss of biodiversity, such as the periwinkle which can be used to cure childhood leukaemia. Large areas of the forest are becoming wastelands as they are not able to replant the forest over large areas	• Little has been done by the government because they need oil revenue • Maxus energy has put an underground pipeline through Yasuni National Park • Maxus energy has built schools for indigenous people • Maxus energy has invested $60m on environmental protection

You could be asked for opinions for and against oil extraction

Remember to learn the specific points

6 Managing hazards

Global distribution of tropical storms

How do you describe the distribution of tropical storms?

In the following example there are three clear statements worth one mark each. You would gain full marks for this answer.

Nearly all tropical storms are found between the tropics of Cancer and Capricorn.

The source regions are all found over the oceans.

Tropical storms are more common in the northern hemisphere than the southern hemisphere.

On the foundation paper students might be asked to fill in gaps in a paragraph. Complete the sentences below.

Choose the words from the following list:

land   ocean   less common   tropics

more common   circles

3 marks

Nearly all tropical storms are found between the tropics of Cancer and Capricorn.

The source regions are all found over the ....

Tropical storms are .... in the northern hemisphere than the southern hemisphere.

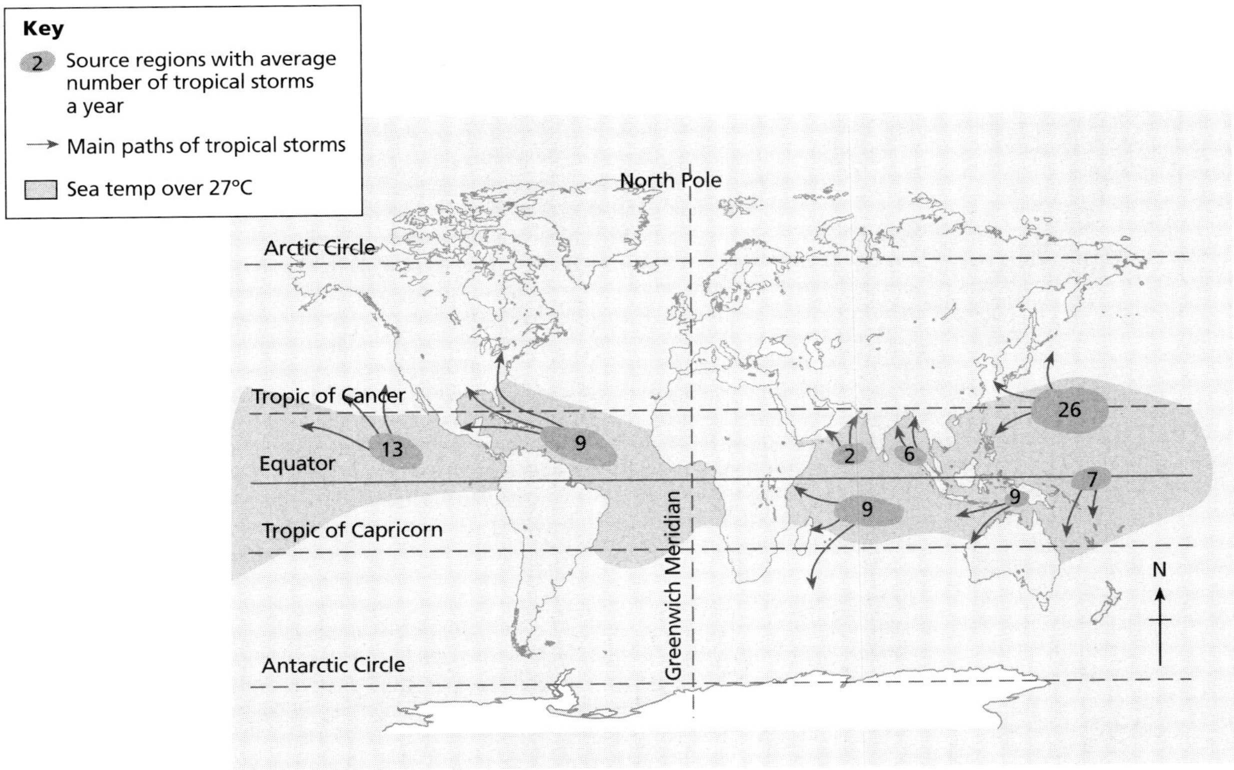

The global distribution of volcanoes and earthquakes

Describe the distribution of volcanoes.

Because the question uses a map it expects you to give specific information from the map.

- Many volcanoes are found around the Pacific Ocean. (For the higher paper you may be expected to know the names for important parts of the world. At foundation level there might well be some place names marked on the map to help the students.)
- There is a line of volcanoes running down the middle of the Atlantic Ocean
- There are no volcanoes in Australia
- Volcanoes are found both on land and sea
- Nearly all the volcanoes on the land are sited close to the oceans

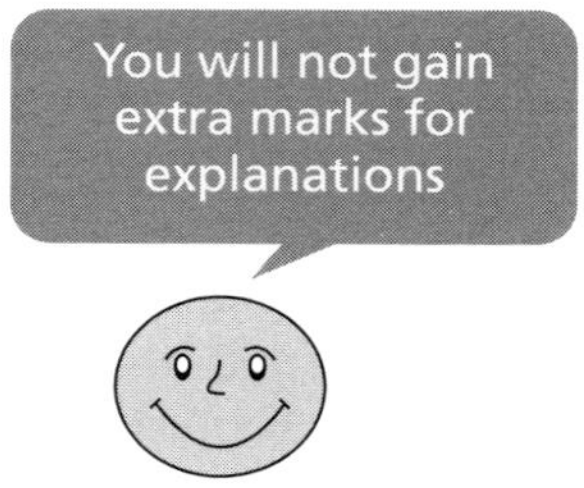

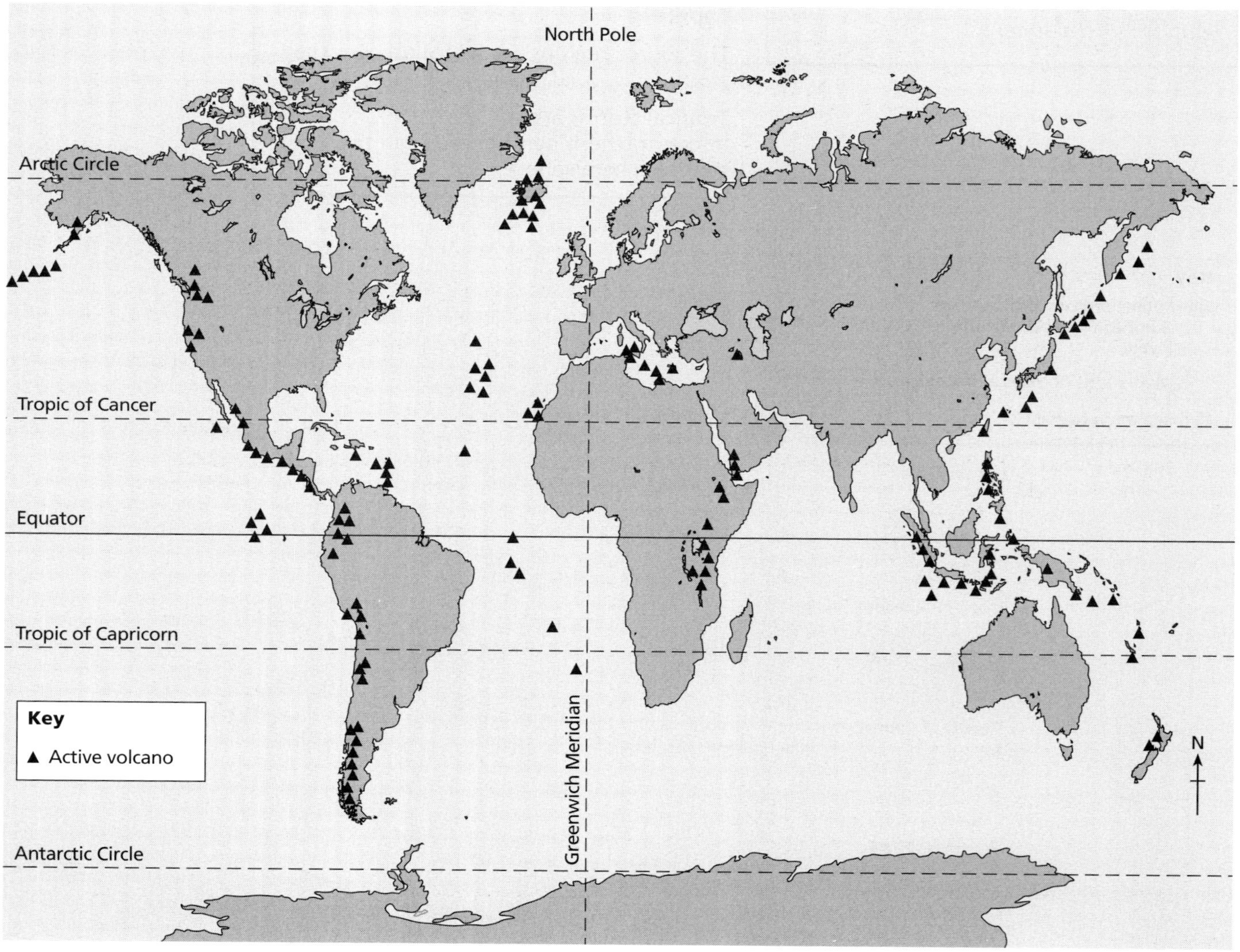

Plate boundaries

The Earth's crust is made up of several tectonic plates. Convection currents in the molten area beneath the crust called the mantle cause the plates to move. **Movement has the greatest impact at plate boundaries where two plates meet.**

There are three different types of plate movement:

- Some plates move towards each other (converge). This is called a **convergent plate boundary**
- Some plates move away from each other (diverge). This is called a **divergent plate boundary**
- Some plates move alongside each other. This is called a **conservative plate boundary**

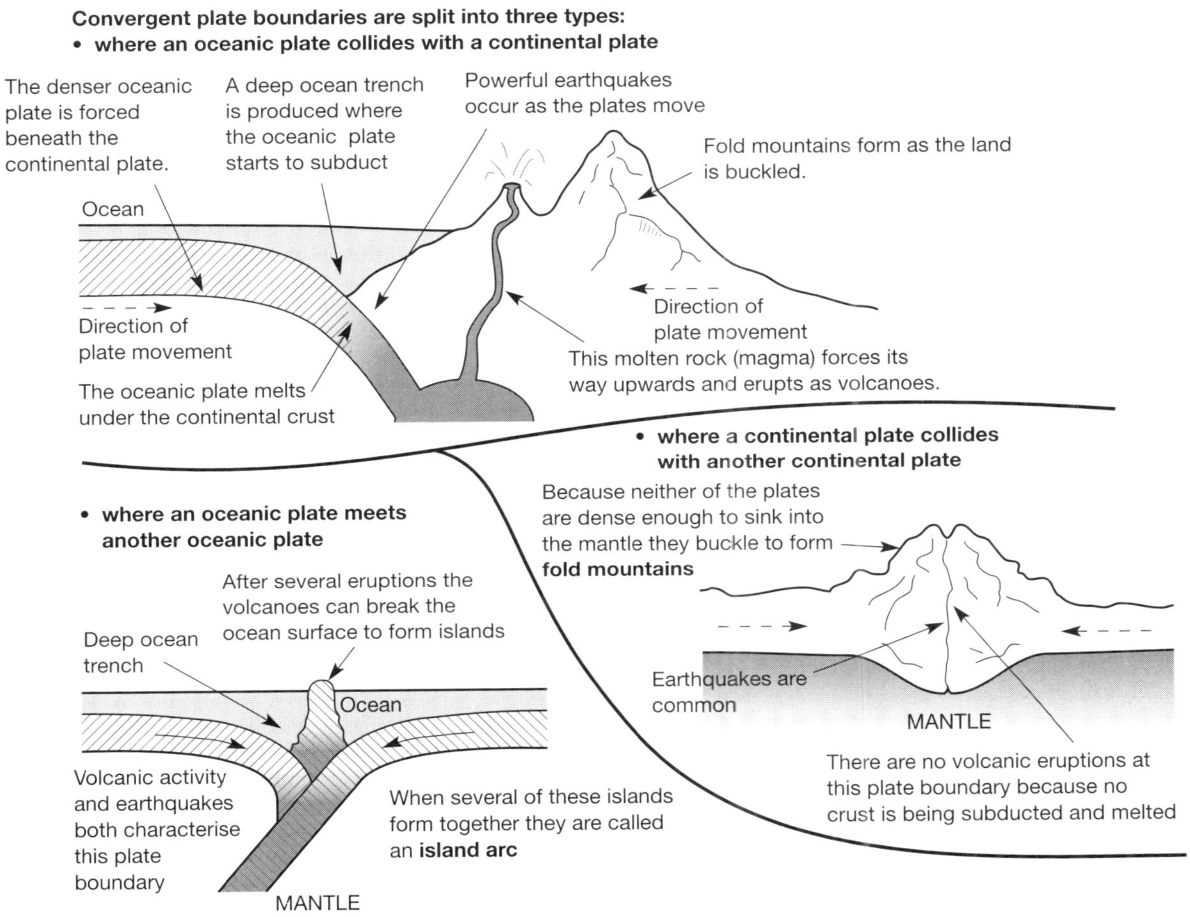

Divergent plate boundary

These are usually found in the middle of oceans.

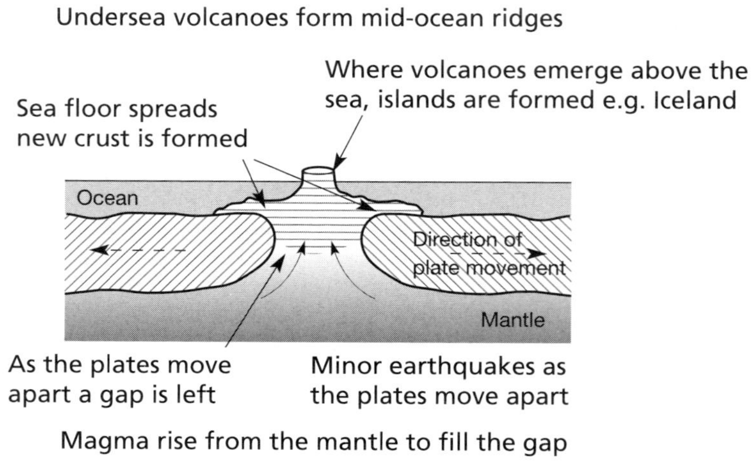

Conservative plate boundary

- Plates move in different directions or at different speeds
- Plates become locked together
- Pressure builds until rock snaps along a fault
- Plates move suddenly causing powerful earthquakes
- No volcanic eruptions happen here

Earthquake

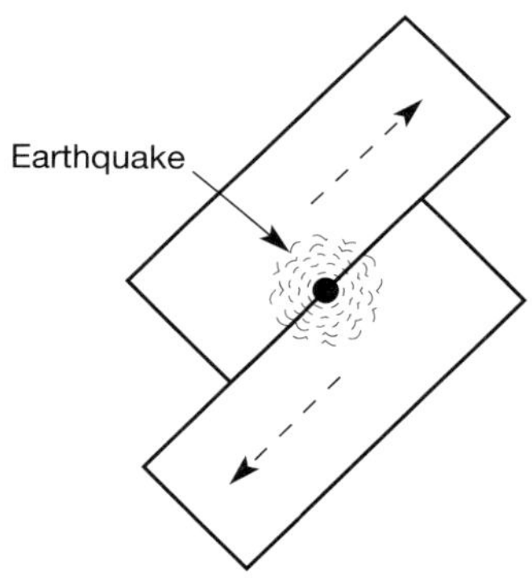

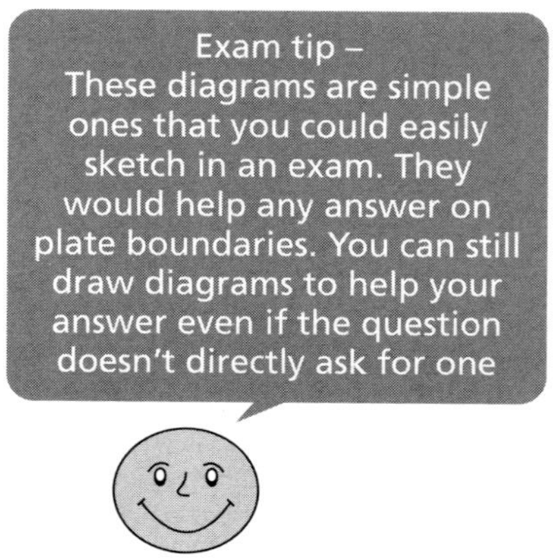

A study of tropical storms

In the examination you might be asked questions on:

- The impact of the storm on people and the environment
- How the development of the country influenced the impact of the storm?
- Why do people still live in areas at risk from tropical storms?

Case Study

Hurricane Floyd a tropical storm in an MEDC

How did Hurricane Floyd affect the people and the environment in southeast USA?

Human impacts

- 79 people died
- 4 million people left their homes in the states of Carolina, Georgia and Florida
- 1 million people lost electricity and water supplies
- nearly 12,000 houses were destroyed or too heavily damaged to repair
- 150,000 people registered for government assistance

Economic impacts

- the total amount of money paid out for insurance: \$460 million
- farmers lost \$1 billion
- many farmers had to find other jobs
- some farmers had their crops totally destroyed
- 250 roads and several bridges were damaged
- in the Bahamas; boats, offices and hotels as well as houses were destroyed

Environmental impacts

- large areas of beaches and sand dunes were destroyed
- Oak Island had 20 metres of its beach blown away
- thousands of trees were blown down
- the Barrier Islands suffered severe erosion which led to an increased flood risk on the coastline of the USA
- 14 states suffered from the effects of flooding. North Carolina alone had an area of 30,000 square kilometres flooded, which is an area the size of Belgium

Case Study

Cyclone One Bravo a tropical storm in an LEDC

How did Cyclone One Bravo affect the people and environment of southeast Bangladesh?

Human impacts

- 111 people died, 7000 injured
- 500,000 people made homeless
- wells were contaminated by sea water
- diseases were common, thousands were affected by diarrhoea
- communication links between Maheshkhali Island and the mainland were severely disrupted

Economic impacts

- 608 schools were damaged
- fishermen lost their incomes as their boats and nets were broken
- 30,000 hectares of land were damaged
- 2000 cows died
- electricity supplies failed in most areas
- roads, bridges and cyclone shelters were destroyed

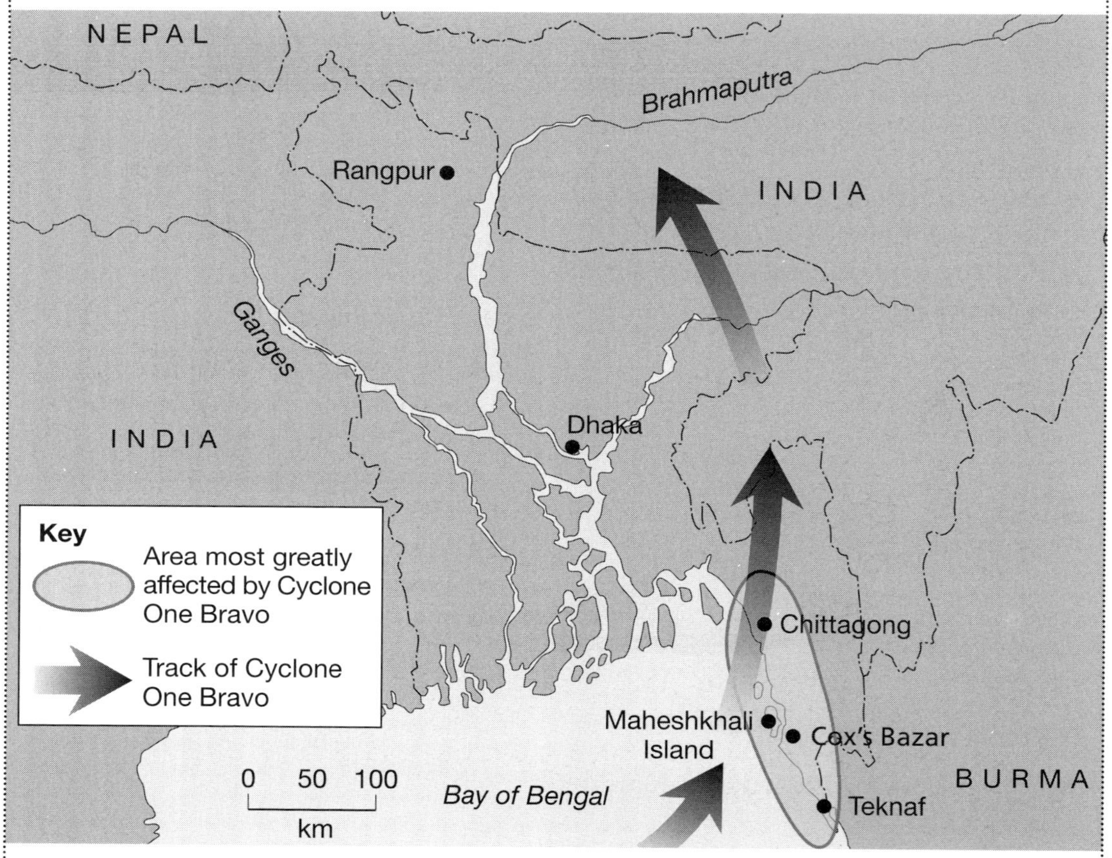

Exam Tip – A* candidates will need to know two or three specific facts for each of the impacts. C grade candidates will need to have some specific knowledge of a case study

Environmental impacts

- fish ponds were contaminated
- flood embankments were damaged leading to increased risk of flooding
- many low lying islands in the Bay of Bengal destroyed
- soil erosion in the foothills of the Himalayas

State of development

How the wealth of a country affects the impact of the storm

Poor country

Bangladesh – LEDC

- $1 billion of international aid
- limited national aid
- no insurance
- poor communication links make cyclone warning difficult
- poorly constructed houses, easily destroyed
- several years for the area to return to normal

Rich country

USA – MEDC

- money available from state and government funds
- most house owners had insurance
- hurricane centre monitors storms
- warnings given out on TV
- hurricane proof buildings built
- area cleared up within a few months

Why do people still live in areas at risk from tropical storms?

LEDC MEDC

Earthquakes – predictions and precautions

Despite much research it is not yet possible to predict earthquakes accurately. Several precautions can be taken.

Earthquake proof buildings – only rich MEDCs have the money to build these.
Emergency drills – Japan have an annual earthquake day where children in schools practise earthquake procedures.
Emergency plans – This is how Californians are told to prepare:

1. have flashlights and batteries ready
2. have first aid kit and fire extinguisher ready
3. store a few gallons of water per family member
4. store one week's food outside the house
5. place beds away from windows and mirrors
6. agree on a plan to reunite the family by day or night

Volcanoes

Predictions

The prediction of volcanic eruptions is becoming much more scientific.

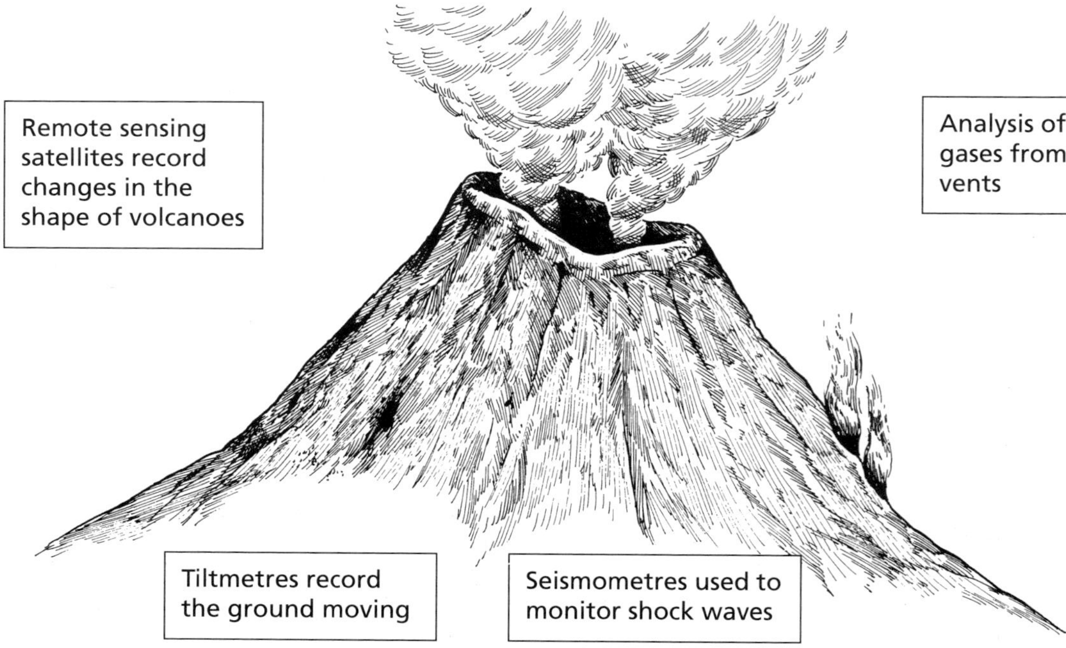

Precautions

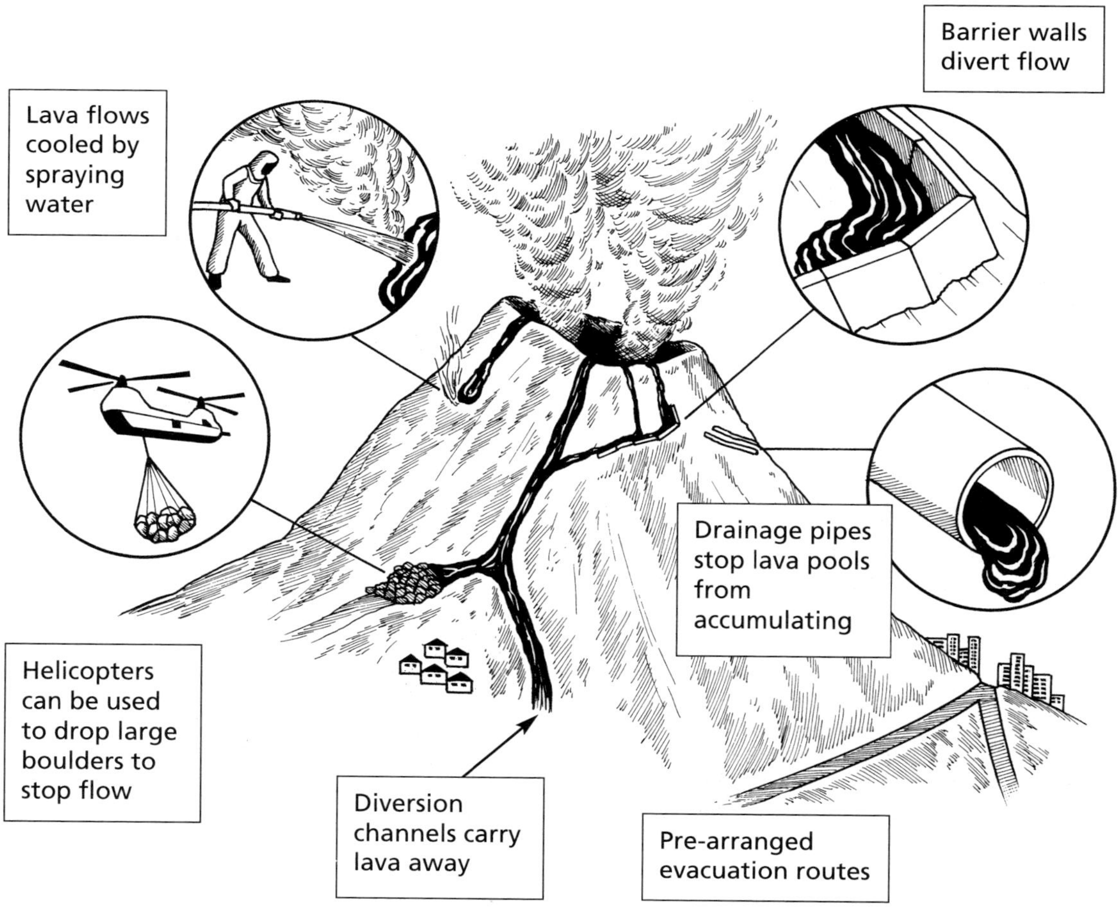

Managing earthquakes and volcanic eruptions

Case Study

LEDC Earthquake Turkey 1999

Short term responses

- Emergency shelters and field hospitals set up by German Red Cross
- Emergency teams with sniffer dogs and heat seeking equipment were brought in
- Water and food supplies organised by the daily *Turkiye* newspaper. The central government only sent bread, which quickly went stale
- Fire fighters putting out fires
- Local inhabitants using their bare hands to search through the rubble for survivors
- American Red Cross donated 25,000 comfort kits and 4 members of its emergency response unit
- Red Crescent presented tents and soup kitchen
- Red Cross established on-line donations

Long term effects and recovery

- Long term effects are considerable
- Homes rebuilt, over 100,000 were destroyed
- Emotional damage to relatives of 2500 dead
- Unemployment until factories and offices rebuilt
- People still living in temporary villages 2 years after earthquake
- Diseases related to unsanitary conditions lasted until hospitals and medical supplies improved

For more info try www. google.com and type Turkey, earthquake, long term recovery

Case Study

MEDC – Volcanic eruption Mount St. Helens 1980

Short term responses

- Monitoring equipment in area checked for further eruptions
- $200 million spent on dredging the Toutle and Columbia Rivers
- Levees built at Kelso
- Evacuation of residents happened quickly
- Emergency equipment and specialists sent to the area to reconstruct Spirit Lake
- Local funds of $60 million immediately available

Long term effects and recovery

- Within 3 years 90% of the original plant species had regrown
- Increased jobs in tourism. 1 million visitors a year
- 3 roads, bridges, and trail bridge rebuilt
- Volcanic deposits increased soil fertility
- President Reagan set up emergency fund of $1 billion, helped long term recovery
- Vegetation slowly recovering 10 million trees replaced

7 Managing tourism

For this unit you need to understand the sectors of industry and the multiplier effect. The causes of the rapid increase in tourism since 1945 and how to classify tourists.

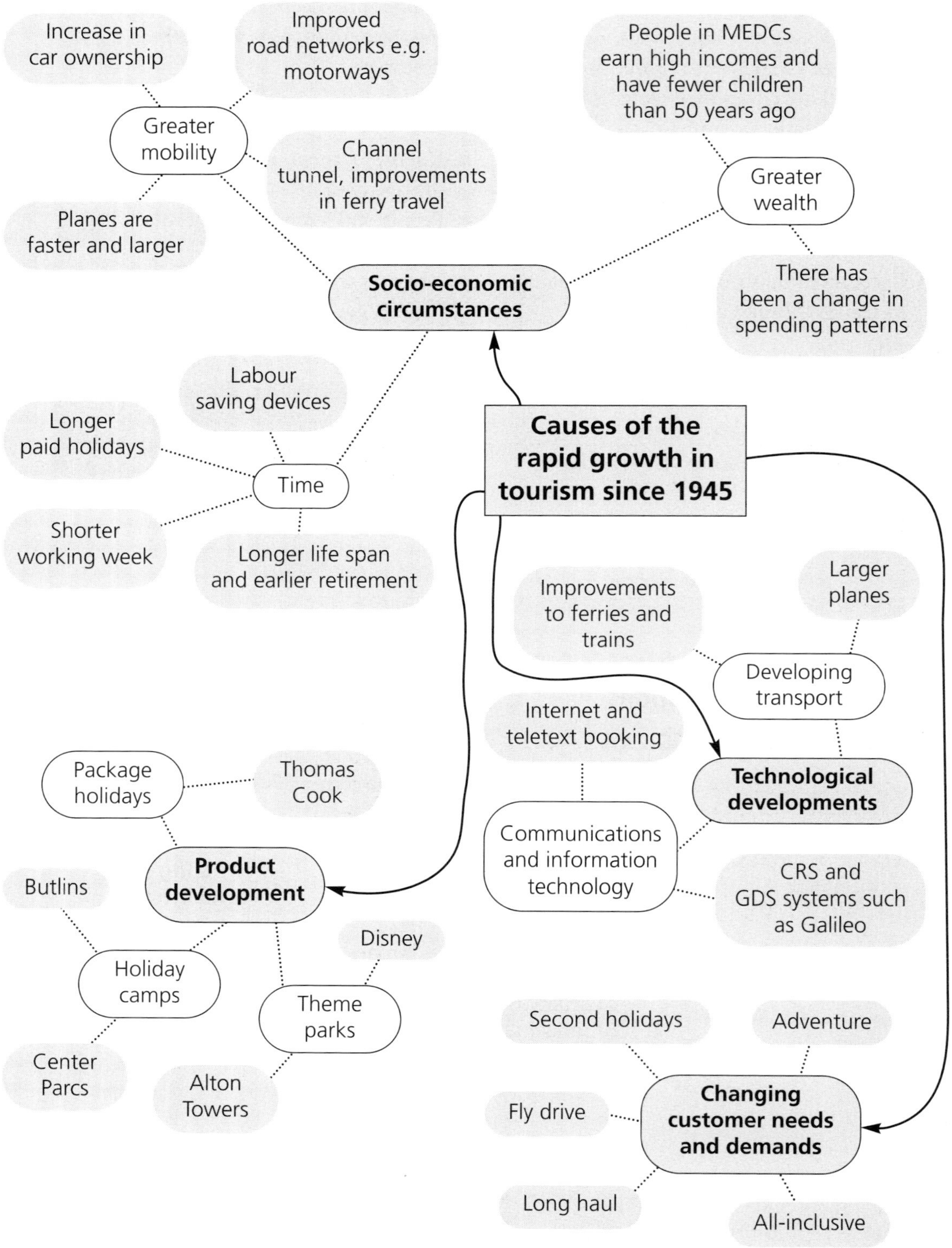

The three sectors of industry are explained in Chapter 3.

Refer back to your notes or do you already know the difference?

Tourism provides many jobs usually in the tertiary sector. When tourism develops in a country it also creates jobs in other sectors due to the multiplier effect. Tourists pay money to stay in a hotel. The hotel owner pays people to clean the rooms. The cleaners then spend the money on food; this gives money to local farmers. Therefore the money spent by tourists is re-spent in the country many times and in a way creates more money.

Classifying tourists

Can this be done?

Hopefully in a number of ways –
Nature of activity, tourists can be either active or passive. Active tourists take part in water sports or skiing, passive tourists go sightseeing or relax on the beach.
Location preference – some tourists like culture others like beach holidays, others mountains.

Other categories are the distance that the tourists have travelled, for example, nationally or internationally. The length or duration of the holiday is it for a weekend, week or fortnight.

Tourism in a coastal area in an LEDC, Zanzibar

The first points you should study are the physical and human attractions.

Physical attractions

1 Cool sea breezes to north and east
2 Annual temperature 28–38°C
3 Scuba diving and snorkelling
4 White sandy beaches at
   1. Matemwe
   2. Uroa
5 27°C sea temperature coral reef

Human attractions

1 Stone Town which is a functioning ancient town with the Palace Museum within its walls
2 Prison Island for unruly slaves which still has the old buildings
3 Spice towns on plantations
4 Chumbe island – private nature reserve
5 Many Mosques e.g. Shirazi Dimbani at Kizimkazi

Now you must learn the **impacts** that tourism has had on Zanzibar.

	Economic impacts	Social impacts	Environmental impacts
Positive	Jobs in hotels at Matemwe. Tourist numbers continue to grow approximately 100,000 by 2002. Tourist money helps the countries balance of payments US$220m a year from tourism.	Improvements to roads which can also be used by locals. New passenger terminal at the port at Stone Town cost $40m; can be used by locals as well as tourists. 1000 locals and tourists pass through the port every day. The development of internet access through cyber cafes which the tourists, especially the many backpackers, demand. These can also be used by locals and have given them many benefits.	Stone town designated as a World Heritage site has helped to protect it as an ancient walled city. Many buildings have been bought by large hotel chains and renovated because of the growing tourist trade. This has helped Stone Town to be preserved, for example, the Zanzibar Serena Hotel. It is tourism that is providing the money to maintain the buildings of Stone Town.
Negative	Loss of fish stock on east coast beaches due to tourists. Loss of access to the beach for locals to fish. Jobs are menial and low paid in hotels. Rise in food prices as locals have to compete with hotels for food. Zanzibar is becoming too reliant on tourism which could cause problems if the tourist industry collapses or Zanzibar goes out of favour with tourists.	Zanzibar is a Muslim country. Tourists from western countries have different moral codes. Tourists are generally wealthy and this has lead to an increase in muggings and robberies on the east coast. Lack of drinking water for locals because of demands from tourists.	Disruption to the breeding ground of turtles. Coral reef around the island is starting to be affected by tourist pressure. Raw sewage is put directly into the Indian Ocean as Zanzibar has no sewage treatment works. At Stone Town tourists walk around the trickles of sewage on the beach.

Now look at the effects on different groups of people.

Below are the views of different groups of people. Remember you should always refer to the group and not write about 'some' people and 'others'.

- village leaders feel that they are not being consulted about the tourist developments
- local villagers who have fished in this area for years are worried about losing access to the beach
- environmentalists are worried about new tourist developments because of the increase pressure on sewage and water supplies
- British backpackers are against large tourist developments because they want the island to remain as it is
- young villagers of Nungwi are happy because they want to mix with people of different cultures
- the government of Zanzibar is happy because more money is coming into the island which will help it to develop.

Tourism in a mountain area in a MEDC, Malham UK

Remember you need to know about the physical and human attractions.

- Malham Tarn
- Malham Cove
- 14°C in July
- 2°C in January
- Janet's Foss Waterfall

Human attractions

- National Park Centre which opened in 1974 provides information on the area
- Pay and display car park
- Area is well marked with signposts and footpaths to the physical attractions
- Many tea shops and public houses for refreshments (The Buck Inn)
- Stone houses which are very old (18th century) and an ancient packhorse bridge

You now need to learn the impacts that tourism has on Malham.

	Economic impacts	Social impacts	Environmental impacts
Positive	Shops are kept viable in the village. Pubs are able to make a profit. Demand created for local services like banks and public transport. Employment opportunities in an area which has high unemployment rates. Local farmers have opened campsites, e.g. Town Head Farm. Housing has become more expensive because of the trend to buy second homes.	Greater provision of services because of tourists creating extra demand, e.g. more bus services in the summer. Shops are open longer in the season because of extra demand from tourists.	Area is being protected because it is a honeypot site. There are a number of management initiatives. • Eroded paths are restored in the Goredale area • Additional footpaths have been created
Negative	Pubs have become themed which is not liked by the locals. Local shop is only open three days a week in the winter because the tourist business is seasonal. Jobs are seasonal.	More cars on narrow lanes, locals find it difficult to go about their daily business. Access is restricted for emergency vehicles because of tourists parking on narrow village streets. Decline of village shop which used to sell groceries for locals and now sells gifts for tourists. Housing has become more expensive because of the trend to buy second homes. This has meant that locals cannot afford them.	Pollution from the many vehicles in the area. Footpaths becoming eroded especially – Janet's Foss waterfall to Goredale Scar as this is closest to the village. Too many billboards put up by businesses. Litter is a continual problem for farmers and local wildlife. Gates are left open and visitors' dogs disturb animals.

But what are the effects on different groups of people?

Below are some views – remember to learn groups of people and both sides of the argument.

Arguments for:
Village newsagent – the extra business coming from tourists has kept the shop open.
Town Head Farm owner – the income from the campsite means that I have not gone bankrupt.
Estate agent – I am earning more money on commission because house prices have risen because of the extra demand.

Arguments against:
English Heritage – the number of tourists at Malham must be controlled; so that it is not destroyed by its own popularity.
Elderly local residents – are annoyed by the extra noise and litter caused by the tourists.
Young married couple – cannot afford a house in the village that they grew up in because people demanding second homes have put up house prices.
Farmer at Friar Farm – tourists walking to Janet's Foss waterfall are always disturbing my animals and leaving litter.

The next section in this unit is about sustainable tourism but what does that mean?

You must learn an example of sustainable tourism in an MEDC and in an LEDC.

Sustainable tourism – this means allowing the development of tourist facilities but in a way that there will be little damage to the environment. In some areas this is too late but steps are being taken to manage the area more sustainably in the future

Sustainable tourism in the Maldives

Factfile on the Maldives

Item	
Location	Indian Ocean, south-west of India
Make up	1190 islands. 280 are inhabited, 80 are tourist resorts
Ownership	The government owns all the islands
Development	1990 the Maldives was classified as one of the 29 least developed countries

Item	
Climate	26–30°C
Rainfall	1900 mm
Tourist development	First resort 1972
Tourist numbers	429,700 in 1999

So how is tourist development on the Maldives being managed to ensure that it is sustainable.

- Resort buildings to be 5m from the vegetation line
- No buildings are allowed above the treetops
- No more than 20% of the island can be built on. This includes water bungalows
- Resorts must have incinerators for their rubbish
- Desalination plants to provide fresh water
- Solar hot water systems are used as much as possible
- Resorts must have their own septic tanks to deal with sewage waste

Remember you may be asked about the issues involved with this

But what does this mean? Does everyone agree with it? Where did the money come from?

Help given to enable the Maldives to achieve sustainable tourism

- Since 1981 – EU given US$3.75m. 1996–2005 Tourist master plan. Help given to train people in customer service, etc.
- Japan – US$14m to build breakwaters along Males shores to help combat rising sea levels
- UNDP & WTO aid in the form of help with resort management and training projects. A school of hotel management and catering has been set up and is affiliated to the Birmingham College of Food, Tourism and Creative Studies
- Norway – scheme to help with the finance of installing incinerators. 1994 – 49 incinerators installed with resorts paying 15% direct to supplier. 42.5% given as grant and 42.5% given as a loan to be paid back over 10 years

The Maldives are islands that have been carefully developed for tourism. Specific islands are designated for tourists and there is little mixing between the western tourists and the Maldivians. This is to preserve the Maldivian culture. What will the Maldivian government need to do to maintain this sustainable management?

Issues/areas for concern

- Coral blasting to provide access to new resort islands has resulted in erosion of island beaches. Breakwaters need to be built to protect these beaches
- Divers and nutrient rich sewage have affected coral reef habitats on some of the first tourist islands. This will need to be carefully monitored on the new islands
- Tourist development will be concentrated in certain areas, e.g. northern atoll of Baa
- There are designated 'home' islands that tourists are allowed to visit to reduce the impact of western cultures. This practice will continue to be strictly monitored
- More local inhabitants will be encouraged to work in the tourist industry
- More local artefacts need to be produced – at present souvenirs are imported from Sri Lanka
- Strict monitoring of existing resorts is needed

Who do you think would be for and against this development?

Arguments for

Environmentalists – the Maldives are being developed in a sensitive way. As long as this continues we see only benefits from this project.
Government of the Maldives – we see tourism as a way of developing the country. However we are not prepared to allow the development of tourism to ruin our islands.
Maldivian businessmen – we are getting help with our projects which means that we can protect our islands as well as develop the tourist industry.

Arguments against

Maldivian people – we spend long periods of time away from our families as we live on the islands with the tourists. We are given very little time off in the season and the pay is not good.
Thomson tour operator – holidays to the Maldives are very expensive because of the management of the islands. This restricts the holidays to people who have a substantial amount of money to spend on holidays.
British family – we cannot afford to go on holiday to the Maldives because the sustainable management has meant that the holidays are very expensive.

Sustainable tourism in the USA, Yosemite Valley

This is your MEDC case study

Yosemite Valley is in Yosemite National Park.

It has many features that people come to see – sheer rock faces such as El Capitan and sub-alpine areas such as Tuolumne meadows. 4m people visited Yosemite in 1995.

There are a number of issues:

Yosemite: the issues!

visitor numbers continue to increase

↓ visitor centre built

↓ camping sites made larger

↓ more car parks

↓ road made one-way but two lanes

↓

visitor numbers continue to increase so that on some weekends the valley is closed

↓

the National Park service needs to bring in more radical reforms to sustainably manage the park or it will be ruined by its own popularity

The plan

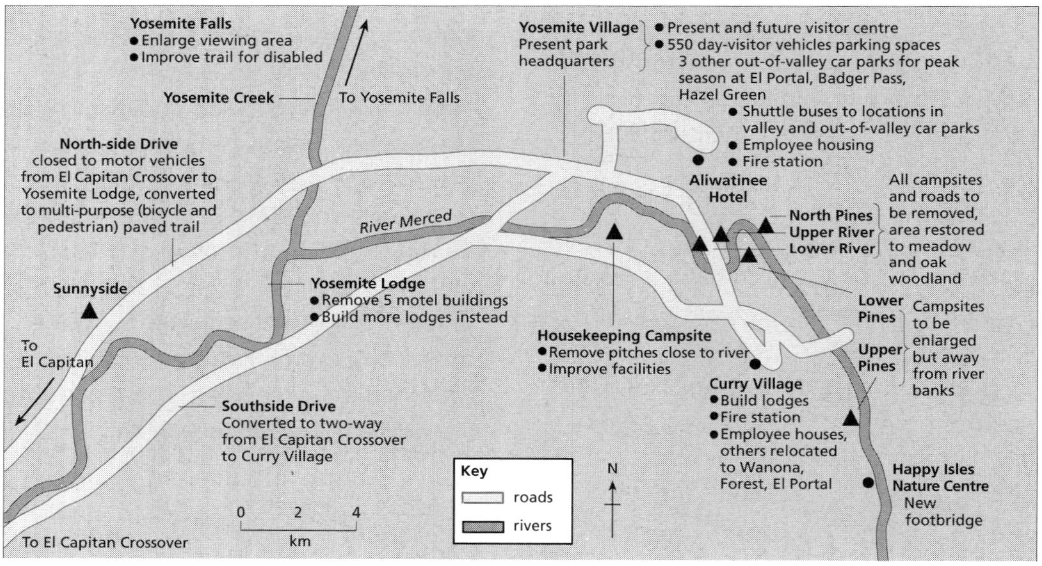

These decisions were taken to protect the valley. However, everyone did not welcome them.

- National park service felt that if they did not act in this way the valley would be destroyed by visitors
- Less physically able visitors felt that they were being penalised because of the lack of access for cars
- Tourists do not like to travel on buses, it is more convenient to use their own cars
- Environmentalists who monitor the levels of pollution from cars in the valley were becoming concerned and welcome the ideas to restrict cars in the valley
- The employees are still being provided with houses but many will have to travel further to their work
- Walkers and cyclists now feel safe on the multi-purpose trail that has been created

8 Managing urban areas

Inner city decline

Redevelopment is when old buildings in a city that are of no use are knocked down and replaced by new buildings.
Renewal is when old buildings are renovated. The outside of the building is often kept the same but the inside is modernised.
Brownfield site is an area within the city, which is no longer used and is available for redevelopment.
Brownfield potential is the number of brownfield sites, which are available for redevelopment within a city.

Outer urban sprawl

Greenbelt is an area around the city, which is made up of non residential or industrial land uses such as farmland and recreation areas. They help to control the spread of urban areas into the countryside.
Greenfield site is an area on the edge of the city, which has never been developed in any way.

You will need to learn these definitions and be able to apply them to a case study that you have learnt. There is a good case study in the Tomorrow's Geography textbook about Reading on pages 172–6. A summary is produced below

How has land use changed in Reading?

1. The inner city

Transport:
- 1969 inner distribution road opened – terrace houses and factories demolished
- 1970 Broad Street was closed to all but essential traffic
- 1990s Friar Street was pedestrianised – decrease in pollution, safe for children

Shopping:
- Oracle project 1997–2000. A large area of the CBD and the old bus garage was demolished and a new shopping centre and cinema complex were built
- Oracle is built on both sides of the River Kennet on land that was wasteland
- 40% of original shops from Broad and Friar Street have relocated to the Oracle
- Friar Street has attracted pubs and restaurants
- Broad Street has become a financial area

Housing:
- Brownfield site in Rose Kiln Lane has seen flats built on the site of derelict factories
- The old Rex cinema site has redeveloped into flats called Lexington Court
- Many more brownfield sites are being developed in Reading to try and meet Berkshires requirement of 67,000 new houses by 2016
- The old football ground at Elm Park has been redeveloped for housing

2. The urban edge

Transport:
- A new dual carriageway road joins the Inner Distribution Road with the M4

Shopping:
- Reading Gate on the southern boundary of Reading is a large out of town shopping centre including B&Q and Comet
- A retail and entertainment area close to junction 12 of the M4 was built in the late 1980s. Sainsbury, Allied Carpets and Superbowl are found here

Housing:
- Beansheaf estate built in the 1980s
- A large modern estate built in the 1990s with mainly 3 and 4 bedroomed houses on the southern edge of Reading

Industry:
- Green Park is a new science park which has been developed next to the M4
- There is a new science park close to the village of Theale.
- The Reading International Business Park has developed close to the new Madejski Stadium

In the exam you might be required to use a map and photographs to show the characteristic features found at the edge of urban areas

Urbanisation

Urbanisation is the increase in the percentage of people living in urban areas.

Urban growth is the increase in the size and total population of urban areas.

Make sure that you know the difference between urbanisation and urban growth

What was the cause of urbanisation in MEDCs?

- Urbanisation in MEDCs began as a result of the Industrial Revolution during the nineteenth century
- People migrated from the countryside to towns looking for work in the factories
- Terraced houses were built close to the factories for the workers to live in
- Over time many of the more favourable towns expanded to become cities
- Urbanisation continued during the twentieth century due to the pull of the 'bright lights'
- 94% of the population of the UK live in urban areas
- Urbanisation in France is 75% and in Australia 85%
- There is now some movement in the UK away from urban areas to more rural areas in search of a better quality of life. This is known as counter-urbanisation

All these land uses pose threats to the Green Belt. Pressure is being put on the Government to allow more development in designated Green Belt areas. Several thousand new homes are to be built in Stevenage in Hertfordshire despite opposition from locals and environmental organisations

What is happening in LEDCs?

- Since the 1950s LEDCs have been experiencing rapid urbanisation

Why?

- There is migration from rural to urban areas because of the lack of jobs and low salaries in rural areas
- There is a high natural increase in population in the cities because of:
  a. high birth rate due to the youth of the migrants
  b. low death rate due to improvements in medical care

In the exam you might be asked to comment on the distribution of large cities in the World. **Remember** with distribution questions write down what you can see. You don't need explanations

Urbanisation in percentages in selected LEDCs

Country	1950	1970	1990	2000
Botswana	0	8	24	53
Brazil	36	56	77	83
Nigeria	10	20	35	43
India	17	20	28	34

Know your case study – Urbanisation in an LEDC

There is an in depth study of São Paulo on pages 166–171 of the textbook. You need to learn this or another example

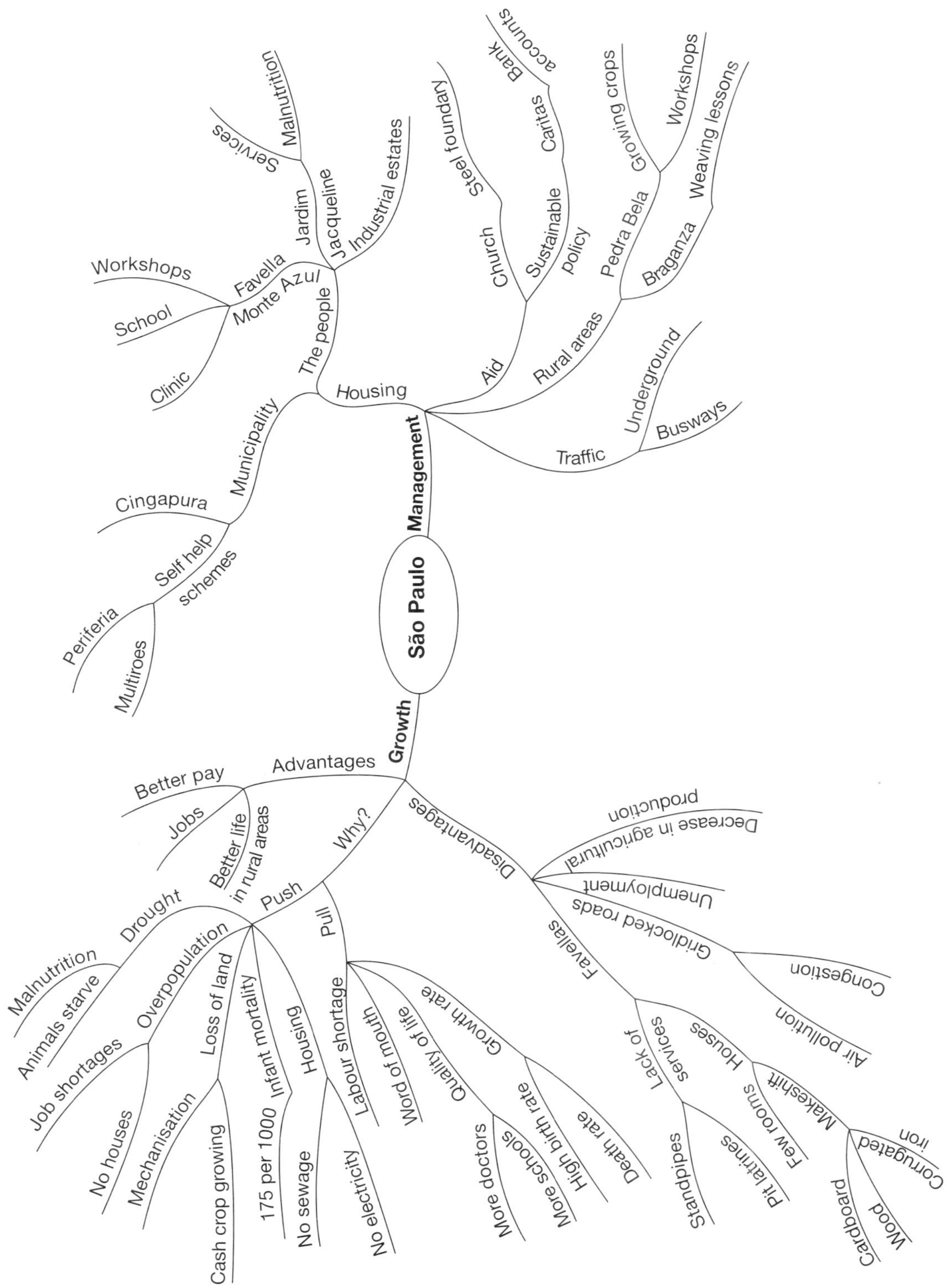

Managing environmental problems in an urban area

This is another case study that you need to know. Tomorrow's Geography has a study on Cairo. This is summarised below. You can use an urban area in an LEDC or an MEDC. You must be able to write about air, noise and land pollution in your chosen urban area

If you want to gain high grades you must include specific facts. It is not good enough to say that cars cause problems. You need to know how many cars and what the problems are. Similarly it is no good to say that pollution causes medical problems, you need to mention the medical conditions e.g. asthma

Causes of air pollution

- 2 million old cars. 60% more than 10 years old. No emission controls
- Heavy industries in the city e.g. iron and steel and cement works
- Many unregistered lead smelters and backyard dumps with no pollution controls
- Primitive solid waste recycling methods
- Sand blown over the city from the desert

Consequences of air pollution

- Kills 10,000 people a year. Lead emissions reduces children's IQ by 4 points. Smoke 3 times higher than recommended levels
- 1 million people live in close proximity to industrial areas. Air pollution is 3 to 4 times above the accepted level
- High lead concentration causes high blood pressure, kidney problems, infertility and nervous disorders
- Unhygienic, many diseases can be caught
- Creates a haze over the city. Particulate material in Cairo is 10 times higher than the World Health Organisation guidelines

Causes of noise pollution

- Cars and constant traffic jams with honking horns and poor exhaust systems
- Constant flow of boats on the Nile
- Clubs and discos on floating nightclubs

Consequences of noise pollution

- Hearing disorders. Bad tempers because of gridlocked roads
- Disrupts the relative peace beside the river
- Annoys the residents by being noisy late at night

Causes of land pollution

- Cairo produces 10,000 tonnes of solid waste each day. Only 60% collected
- Zaballeen only collect waste from residents who can pay
- Hazardous wastes from chemical plants are not disposed of safely
- Medical waste dumped with ordinary waste

Consequences of land pollution

- 40% of waste thrown away to rot. Rats thrive and spread disease
- Poorest people dump their rubbish
- 50,000 tonnes of toxic waste has stockpiled
- Can lead to the spread of infectious diseases

What sustainable strategies have been developed?

- 1999 air quality monitoring was introduced
- An air improvement project has been set up
- Schemes are in place to reduce pollution from lead smelters
- Vehicle emission testing has begun
- Buses being switched to cleaner natural gas fuel
- Underground expanded, which will hopefully take more cars off the road